Extreme Habitable Environments

A Bridge between Astrophysics and Astrobiology

Madhu Kashyap Jagadeesh and Usha Shekhar

Department of Physics
Jyoti Nivas College
Bengaluru, India

CRC Press is an imprint of the
Taylor & Francis Group, an **informa** business
A SCIENCE PUBLISHERS BOOK

Cover credit: ESA/Hubble, M. Kornmesser

First edition published 2022
by CRC Press
6000 Broken Sound Parkway NW, Suite 300, Boca Raton, FL 33487-2742

and by CRC Press
4 Park Square, Milton Park, Abingdon, Oxon, OX14 4RN

CRC Press is an imprint of Taylor & Francis Group, LLC

Library of Congress Cataloging-in-Publication Data (applied for)

ISBN: 978-0-367-25526-8 (hbk)
ISBN: 978-1-032-25135-6 (pbk)
ISBN: 978-0-429-28959-0 (ebk)

DOI: 10.1201/9780429289590

Typeset in Palatino Lt Std
by TVH Scan

Dedication

The authors humbly dedicate this book

To

Mr. Jagadeesh and Mrs. Sathyavathi
(Parents of primary author)

Late Mr. Shekhar
(Husband of co-author)

Foreword

Astrobiology is a relatively new field of science dedicated to the question of origin, evolution and, ultimately, a place of life in the Universe. It is an interdisciplinary field as it involves biology, chemistry, geology, planetology, even ecology, as well as necessarily recently emerged subjects such as Deep Learning, Machine Learning and Artificial Intelligence. Astronomy, on the other hand, is probably the oldest science on Earth. But astrobiology, or rather it's ultimate question about Life, cannot be answered outside astronomy, as life is truly the product of the Universe, just like everything else around us. This book is a brave attempt to connect these subjects and to unify them.

Teaching astronomy is often difficult and maybe even frustrating because of basic student misunderstanding of major astronomical ideas and concepts. This book is written by the hard-core practicing astrophysicists, who not only perform research but also teach astronomy at basic levels. Thus, the authors understand the problems that students (or a general public) may have in understanding the fundamentals and that is why they tried to explain everything in interesting details. This book may be useful to astronomers, who are trying to understand the astrobiology – its biological nuances, and to (astro) biologists, connecting the biology, or life, ultimately to the astronomical Universe, to its origin and evolution.

Margarita Safonova

Preface

This book provides the foundation material necessary for studying astrobiology by students with knowledge of Physics and Biology.

The book is presented in three sections: the first section deals with the basics of astronomy and astrophysics, the second section is an introduction to the solar system planets and the third section, which is the focus of the book, gives an introduction to Exoplanets. Chapters 4 to 9 in section 3 lay a strong foundation to the various aspects of Exoplanets. Chapter 10 deals with the mathematical bridge between astrophysics and astrobiology. The final chapter 11 gives updates on current research in the field of astrobiology.

Each chapter carries images and diagrams relevant to the topics. Where essential, mathematical problems have been solved as worked examples and several exercise problems are included as assignments for the students to hone up their problem solving skills.

The authors have done their best to present the contents of this book in a manner catering to the needs of students desirous of pursuing their interest in the field of Exoplanets.

Acknowledgements

We would like to thank Dr. Sr Elizabeth (late), former principal and Dr. Sr Lalitha Thomas, the Principal of Jyoti Nivas College, Bengaluru, Karnataka, India for their encouragement and support. We would also like to thank the staff of the Department of Physics of Jyoti Nivas College for their cooperation. We would like to express our special thanks to Dr. Ravi Margasahayam from NASA for his inspirational words. The primary author wishes to thank Dr. Margarita Safonova, former visiting scientist from the Indian Institute of Astrophysics, Bengaluru for her constant guidance and unrelenting assistance. We express our sincere appreciation to Mrs. Vani Kashyap, Kumari. Anvi Kashyap, Mr. Karthik Jagadeesh, Mr. Prashanth and Ms. Varsha Kari for their understanding and wholehearted endorsement without which this book would not have been possible.

The authors wish to gratefully acknowledge Ms. CRC Press, Taylor and Francis Group, Florida, USA for undertaking the publishing of this book.

This book started taking its shape during the beginning of COVID-19 pandemic. We offer our heartfelt salutations to all the frontline workers who risked their lives to keep everyone safe.

We invite constructive suggestions from our well-wishers for further improvement of the contents of the book.

Madhu Kashyap Jagadeesh
Usha Shekhar

Contents

CHAPTER 1

Basic Astronomy

'Curiosity is one of the permanent and certain characteristics of a vigorous mind'

– The Rambler (1751)

The word *'astronomy'* owes its origin to the Greek words - *'astron'* meaning *'star'* and *'nomy'* meaning *'law'* or *'culture'*. As per dictionary, astronomy is *"The study of objects and matter outside the Earth's atmosphere and of their physical and chemical properties"*. Astronomy is one of the oldest natural sciences with a rich history. In modern times, however, astronomy has become synonymous with astrophysics (Unsöld and Baschek 2002). Astronomy is also the latest science that keeps evolving! Exciting discoveries are being made everyday and sophisticated instruments and techniques are being developed to let us peer back into the more distant past. Dedicated amateurs can still make significant contributions to this 'ever-green' branch of science.

1.1 Introduction to Stars

On a clear, moonless night, away from the glare of city lights, you can see with naked eyes a milky patch in the sky interspersed with a myriad of tiny and bright twinkling stars. This milky patch is aptly called the Milky Way, the galaxy in which our solar system is situated. There are an estimated 100 billion galaxies in the observable universe and our sun is just a medium star amongst the 100 billion stars in the Milky Way. As you gaze at the stars, you may wonder: What is the pattern or purpose of the starry heavens? You are not alone in asking these questions. The beauty and mystery of space have always fascinated man.

Stars are the most exciting objects in the cosmos since they have been shining brightly ever since they were born billions of years ago, and they are pivotal for scaling the universe: from smaller celestial objects such as planets, asteroids and comets, right up to large structures such as galaxies, open clusters and globular clusters. A star is basically an enormous spherical ball of hot plasma in which gravity tends to pull the stellar matter inwards, while the energy released during thermonuclear fusion reactions tend to push everything radially outwards; this strikes out a balance between gravity and radiation pressure called *hydrostatic equilibrium*. From a study of the structure and evolution of stars, it is evident that the basic observed quantities like mass, luminosity, radius and chemical composition of a star remain constant almost throughout its life, implying that stars in hydrostatic equilibrium are stable.

1.2 Internal Structure of a Star

Hydrostatic Equilibrium

A star is born due to gravitational contraction of an initially enormous, highly distended cloud of gas and dust particles. During contraction, the entire stellar matter races towards a common center till a stable star is formed. This star will have maximum density and temperature at its core. Density decreases radially as we move from the core to the surface of the star. As per the linear density model, it is assumed that density decreases linearly with radial distance. If ρ_c is the density at the center of a star of radius R and ρ_r is density at a radial distance 'r', then

$$\rho_r = \rho_c(1 - r/R)$$

As per the linear density model, a stable star can be imagined as a conglomeration of thin spherical shells of stellar matter from the densest at the core to the rarest at the surface. Consider the equilibrium of one such shell of radius 'r' and thickness 'dr' as shown in Figure 1.1. Let $\rho(r)$ be the density of stellar matter in that shell and $dm = 4\pi r^2 \rho(r)dr$ be the mass of matter contained in the elementary shell. Let $m(r)$ be the mass of stellar matter contained in the sphere of radius 'r'. The inward gravitational force on matter in the elementary shell is given by

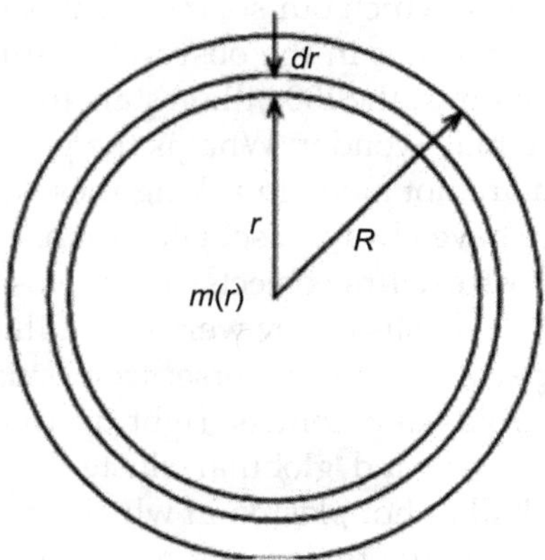

FIGURE 1.1 Mass distribution of a star in spherical shells

$$dFg = Gm(r)dm/r^2 = Gm(r)\rho(r)4\pi r^2 dr/r^2 = 4\pi Gm(r)\rho(r)dr$$

where, G is the universal gravitational constant.

$$\text{Inward gravitational pressure} = 4\pi Gm(r)\rho(r)dr/4\pi r^2$$
$$= Gm(r)\rho(r)dr/r^2 \quad (1.1)$$

Let P and $(P + dP)$ be the pressure (due to thermal pressure of stellar gas) acting on the lower and upper surfaces of the elementary shell.

Net outward gas pressure acting on matter within the elementary shell is $P - (P + dP) = -dp$.

Negative sign indicates that gas pressure increases with depth.

Force on elementary area due to gas pressure is

$$dF_P = -(\text{area of elementary shell})\,dP = -4\pi r^2 dp \quad (1.2)$$

A star attains mechanical equilibrium when the inward gravitational force on each of its elementary shells is balanced by the outward gas pressure.

i.e. when $$dF_P = dF_G$$

Condition for hydrostatic equilibrium is obtained by equating (1.1) and (1.2):

$$dP/dr = -\,Gm(r)\rho(r)\,/r^2 \quad (1.3)$$

Equation (1.3) is referred to as the equation of hydrostatic equilibrium which is also known as the pressure gradient of a star (Karttunen et al. 2007). All the main sequence stars in the *H-R* diagram are in a state of hydrostatic equilibrium.

Equation (1.3) can be used to arrive at the core temperature and pressure of a star. Central temperature of the Sun is around 25 million kelvin and the corresponding pressure is around 26 petapascal (PPa), i.e. 26×10^{15} pascal!

1.3 Energy Production in Stars

The Sun and other main sequence stars have been radiating enormous amounts of energy at nearly constant rates for billions of years. Spectral analysis of star light reveals that most main sequence stars are composed of 90% hydrogen, 9% helium and 1% of other elements. It is now well known that thermonuclear fusion of lighter nuclides to heavier ones is what powers a main sequence star. Typically required temperature and pressure for a thermonuclear reaction are naturally present in the interior of these stars. The actual thermonuclear mechanism or production of energy in a star depends on its mass at birth. In low-mass stars like the Sun, the core temperature does not go beyond 10 to 20 million kelvin. The suggested fusion reaction is the proton-proton cycle or the *p-p* cycle. It is a three-step process: initially, two hydrogen nuclei, i.e. two protons fuse together to form a deuteron, a heavier isotope of hydrogen. During the fusion, a positron, i.e. the antiparticle of electron along with a massless, chargeless, particle called 'neutrino' moving with the speed of light, is given out, according to the equation

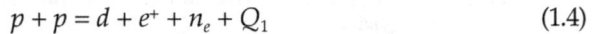

$$p + p = d + e^{+} + n_e + Q_1 \tag{1.4}$$

The deuteron so formed fuses with another proton to form helium 3, a lighter isotope of helium, simultaneously releasing gamma ray photon according to the following equation:

$$d + p = \mathrm{He}_3 + Q_2 \tag{1.5}$$

Two helium-3 nuclei again fuse together to form a helium-4 nucleus in accordance with the equation

$$\mathrm{He}_3 + \mathrm{He}_3 = \mathrm{He}_4 + 2p + Q_3 \tag{1.6}$$

Reactions represented by equations (1.4), (1.5) and (1.6) stand for the proton-proton cycle (refer Figure 1.2) and can be denoted by the single equation

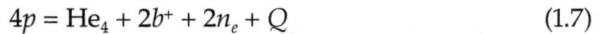

$$4p = \mathrm{He}_4 + 2b^{+} + 2n_e + Q \tag{1.7}$$

Here, $Q = Q_1 + Q_2 + Q_3$ stands for the amount of energy released during the fusion of four hydrogen nuclei into a helium nucleus. Q has to be staggeringly large to sustain the incessant burning of a star throughout its life.

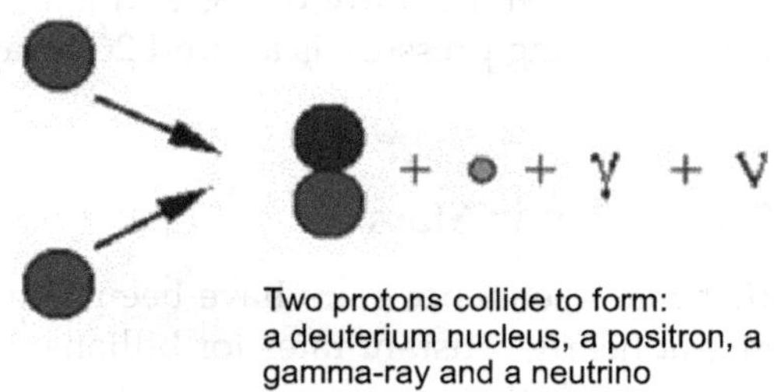

+ • + γ + ν
Two protons collide to form:
a deuterium nucleus, a positron, a
gamma-ray and a neutrino

FIGURE 1.2 Proton-proton cycle (Sarang 2016)

As per Albert Einstein's famous mass-energy equivalence, the source of energy Q in equation (1.7) is the loss in mass 'm' of the reactants during fusion, which is released as energy as per the equation $E = mc^2$.

Taking proton mass = 1.007825 amu, mass of helium nucleus = 4.002603 amu and electron mass = 0.000549 amu, it follows from equation (1.7) that the total mass of the reactants = $4 \times 1.007825 = 4.031300$ amu, total mass of products = $4.002603 + 2 \times 0.000549 = 4.003701$ amu, and loss in mass during each fusion reaction = m = 0.027599 amu. Since 1 amu of mass is equivalent to 931 MeV, $Q = 0.027599 \times 931 = 25.69$ MeV. (amu stands for Atomic Mass Unit which is one twelfth the mass of an atom of Carbon-12 = 1.661×10^{-27} kg).

The Sun, for example, converts 564 million tons of hydrogen every second into 560 million tons of helium. The difference of 4 million tons is converted into energy, which the sun radiates as heat, light and also X-rays, ultraviolet rays and radio waves. This way the Sun loses about 10^{-7} of its mass over a million years. The above reaction is commonly known as PP I chain reaction and 91% of energy produced in the sun is due to this reaction. PP I reaction is dominant in the core temperature range of 10 to 14 million K. Also, 9% PP II (dominant at temperatures of 14-23 million K, involving lithium burning) and 0.01% PP III chain reactions at temperatures in excess of 23 MK are observed as additional mechanisms of energy production (Karttunen et al. 2007). In 1938, Weizsäcker and Bethe suggested an alternative thermonuclear reaction for the fusion of hydrogen into helium, with carbon and nitrogen acting as catalysts. This carbon-nitrogen cycle or CNO cycle or CN cycle is used to explain the energy production of stars that are more massive than our Sun (Freedman and Kaufmann 2007).

The probability of occurrence of CN cycle in a star depends on its temperature and is more likely in low temperature stars. It is believed that in our middle aged Sun, the *p-p* and the C-N cycles could be taking place with equal probabilities. When the *p-p* and CNO cycles exhaust the hydrogen fuel at the core of a star, gravitational contraction supplies the energy needed to keep the star glowing. This contraction continues till the core density reaches around 10^8 kgm^{-3} and the corresponding core temperature is around 100 million kelvin. Austrian Astrophysicist Edwin Ernest Salpeter suggested that at this temperature fusion of 3 helium nuclei, i.e. 3-alpha particles to a carbon nucleus, is possible through the triple alpha or 3α process. The famous triple alpha process takes place in stellar interiors at high temperatures with helium abundance. The following 3-alpha reactions are supposed to take place in the red giant stage of a medium-mass main sequence star:

$${}_2He^4 + {}_2He^4 + 95 \text{ keV} = {}_4Be^8 + \gamma$$

The above reaction is endothermic. ${}_4Be^8$ is unstable and decays with a mean life of 10^{-14} second into 2-alpha particles according to the equation

$$_4Be^8 = {}_2He^4 + {}_6C^{12} + \gamma + 7.4 \text{ MeV}$$

In supermassive stars, where the core temperature reaches as high as 10^9 to 10^{10} K, fusion of carbon, oxygen and silicon continues in sequence till a stellar core of iron is formed (Karttunen et al. 2007).

1.4 Luminosity and Size of a Star

Although the stars in the sky appear to be point sources of light, they actually are of different sizes, i.e. radii, and radiating different amounts of energy at different rates. Total energy radiated by a star in unit time across the entire wavelength range is known as its Bolometric Luminosity or intrinsic brightness. From Stefan-Boltzmann law, power radiated by unit surface area of a star is directly proportional to the 4th power of its absolute temperature (T). If R is the radius of a star of luminosity L, then

$L = 4\pi R^2 \sigma T^4$, where σ is Stefan-Boltzmann constant $= 5.67 \times 10^{-8}$ m^{-2} K^{-4}

The luminosity of a star depends on its radius and surface temperature. The Sun with a radius of 6.96×10^5 km and a surface temperature of 5770 K has a luminosity of 3.83×10^{26} Watt.

1.5 Brightness of a Star

Apparent Magnitude of a Star

A star is visible because it radiates energy at least partly in the visible region of the electromagnetic spectrum, i.e. in the wavelength range of 400 to 800 nm. How bright a star appears depends on the amount of light falling on the retina of the observer's eye. However, classifying the stars based on naked eye observation can be misleading as it does not directly relate to the actual luminosity of the star. Luminosity of a star is its intrinsic property. It is the rate at which energy is radiated into space every second by the surface of the star. It is measured in Js^{-1} or Watt. A very bright star far away from the Earth may seem dimmer than a fainter star closeby. Apparent brightness of a star is the rate at which energy from the star is received by unit area of the earth. It is measured in Wm^{-2}. Apparent brightness (b) indicates how bright a star appears for an observer on Earth and it depends not only on its luminosity (L) but also on its distance (d) from the Earth.

$b \propto L/d^2$ or $b = L/4\pi d^2$, where $1/4\pi$ is the constant of proportionality.

Hipparchus, a Greek philosopher of the 2nd century *BC*, first proposed the apparent magnitude scale for the measurement of apparent brightness of stars. From naked eye observations, it was established that the brightest stars were a hundred times as bright as the dimmest stars. Hipparchus assigned an apparent visual magnitude (*m*) of 1 to the brightest stars and classified the dimmest stars as of 6th magnitude. The other stars were allotted '*m*' values between 1 and 6. It was also noticed that two stars belonging to two adjacent magnitudes have the same brightness ratio. If $b_1, b_2, b_3, \ldots$ etc. are the apparent brightness of stars of apparent magnitude 1, 2, 3, … etc., then $b_1/b_6 = 100$. Also, $b_1/b_2 = b_2/b_3 = b_3/b_4 \ldots$ constant, say A, $b_1/b_6 = A^5 = 100$ or $A = 2.512$, i.e. every star is around two and half times as bright as a star belonging to the next higher apparent magnitude.

Modern astronomers use a color filter and a photocell for measuring the apparent brightness of a star and the currently used photovisual magnitude scale still tallies with the Hipparchus scale. In 1856, Norman Robert Pogson, an English astronomer noticed that the response of the human eye is proportional to the logarithm of the rate at which it receives radiant energy, i.e. if the eye registers a linear increase in the brightness of stars, a photometer records a geometrical increase in brightness. Accordingly, Pogson proposed a mathematical relation connecting the apparent magnitude '*m*' of a star and its apparent brightness '*b*' as:

$$m = C - 2.5 \log b \tag{1.8}$$

where C is a constant that decides the 'zero' of the magnitude scale. The star Vega radiating at the rate of 2.52×10^{-8} Wm^{-2} is said to be of 'zero' magnitude. Stars brighter than the 1st magnitude stars detected with optical telescopes are assigned 'zero' and negative '*m*' values. For example, the apparent magnitude of the blazing Sun is –26.83! Faintest stars observable with the 200" telescope at Mount Palomar, California have apparent magnitude ranging from +22 to +24.

1.6 Distance Measurement of Stars

Distance in astronomy is measured in units that are very large compared to the meter. Mean Earth or Sun distance, called the Astronomical Unit (AU), is taken as the standard for specifying distances of nearby celestial objects (1 AU = 1.5×10^{11} m). Larger distances are measured in terms of Light Year (ly), which is the distance covered by light in one year with its speed of 3×10^8 ms^{-1} in vacuum. 1 ly = 9.5×10^{15} m = 6.3×10^4 AU. Very large distances are measured in parsec (pc), a short form of parallax-second and 1 pc = 3.26 ly. The Sun is at a mere distance of 1.58×10^{-5} light years. Light from the Sun takes 8.3 minutes to reach the Earth. The Sun's distance from Earth can also be expressed as 8.3 light-minutes.

The Earth, while spinning about its axis, also orbits the sun. For an observer on Earth, the nearby stars seem to change their positions against the background of more distant stars. This is similar to stretching your hand in front of you with the thumb erect. As you alternately close your eyes and watch the thumb, it seems to coincide with different objects in the background. This apparent change in the position of an object when viewed from two different angles is called parallax. If you hold the thumb closer to you and repeat your observation, the parallax will be more. It is the same with stellar parallax as well. Parallax is more for nearby stars than distant stars and hence is an indicator of the distance of the star from the Earth. Stellar parallax is measured as half the apparent angular shift in the position of a star as observed from the Earth from two different positions along its orbit around the Sun that are six months apart.

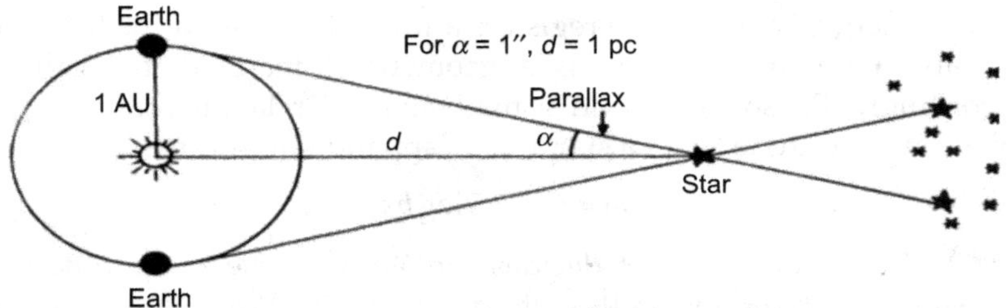

FIGURE 1.3 Geometrical parallax method of distance determination of stars (Credit: Agarwal)

According to Figure 1.3, α is the parallax of a nearby star at a distance d from the sun and it is half the angle between the lines of sight six months apart as marked in the above figure.

From the Figure 1.3, $\tan \alpha = 1$ AU/d or $d = 1$ AU/α (here $\tan \alpha \sim \alpha$).

Distance of a star, which has a parallax of 1 arcsecond, is said to be at a distance of 1 parsec (pc).

$$1 \text{ AU} = d \text{ (in pc)} \times \alpha \text{ (in arcsec)}$$

For 1 arcsec, 1 pc = 1 AU/1 arcsec = 1 AU/(1/60) × (1/60) × (1/57.2958) = 206265 AU = 3.26 ly = 3.084×10^{16} m.

Kilo-parsec (kpc = 10^3 pc) and Megaparsec (Mpc = 10^6 pc) are used to express distances of more distant sky objects. Example: Proxima Centauri, the star closest to the Earth, is at a distance of 1.3 pc and has a parallax of 0.763″.

Absolute Magnitude of a Star

On a clear cloudless night, a person with normal vision can see around 2000 to 3000 stars of varying brightness across the sky. Absolute magnitude of a star is a measure of its absolute brightness or intrinsic brightness, the rate at which it radiates energy from unit area of its surface. In order to compare the absolute brightness of stars, they are assumed to be at the same distance of 10 pc or 32.6 light years from the Earth.

The star's absolute magnitude in terms of its absolute brightness B is

$$M = C - 2.5 \log B \tag{1.9}$$

$$B = L/4\pi D^2$$

where $D = 10$ pc.

$(m - M)$, the difference between the apparent and absolute magnitudes of a star, is known as its ***distance modulus relation***.

From Eq. (1.8) and (1.9), we get

$$(m - M) = 5 \log d - 5$$

or

$$d = 10^{0.2\,(m - M + 5)} \tag{1.10}$$

This equation gives the distance of a star from the Earth in parsec. It is the distance of a star that determines the relation between its apparent and absolute magnitudes. In practice, apparent magnitude is measured using a light meter attached to the viewing telescope and the absolute magnitude is measured using a bolometer.

Case 1: When $d = 10$ pc, $m = M$, i.e. a star has the same apparent and absolute magnitudes if it is at a distance of 10 parsec from the Earth.

Case 2: When $d > 10$ pc, $m > M$, i.e. stars farther than 10 pc appear dimmer than they actually are.

Case 3: When $d < 10$ pc, m $< M$, i.e. stars closer than 10 pc appear brighter than they actually are.

The Sun, for example, has an apparent magnitude of –26.83, while its absolute magnitude is a mere +4.72. A star radiates similar to a blackbody spanning the entire electromagnetic spectrum. Bolometric magnitude gives the absolute magnitude of a star, taking into consideration all the wavelengths in which it radiates. Visual magnitude gives the absolute magnitude of a star only in the visible range of 400 to 800 nm.

1.7 Color Index of Stars

In the early days of astronomy, apparent magnitudes of stars were determined on the basis of brightness ratio alone. With the advent of the photoelectric photometer, more precise measurement of stellar magnitude became possible. In 1953, Johnson and Morgan established the UBV magnitude system for arriving at the absolute magnitude of a star and hence its surface temperature. UBV is the most popular technique for color-temperature determination of a star. *U*, *B* and *V* are bandpass filters in the ultraviolet (*U*), blue or photographic (*B*) and visual or green-yellow (*V*) regions with effective bandwidths of 680 Å, 980 Å and 890 Å centered around 3650 Å, 4400 Å and 5500 Å, respectively. UBV scale is so defined that for an A0V star, $U = B = V$ and $(U - B) = (B - V) = 0$.

Color index (CI) of a star is defined as the difference between the magnitude of a star in two wavelength bands and is indicative of the degree of redness or blueness, i.e. degree of hotness or the surface temperature of the star and hence its spectral type. Prior to 1950, CI was considered for photographic and photovisual magnitudes. The UBV system provides two color indexes, namely $(U - B)$ and $(B - V)$, though $(B - V)$ color index is more commonly used.

$$(B - V) = m_B - m_V = M_B - M_V$$

and

$$(U - B) = m_U - m_B = M_U - M_B$$

Both the CI are zero for an A0V star, negative for blue stars and positive for red stars. Using these CI, the color ratios b_V/b_B and b_B/b_U are determined, which are pointers to the surface temperature of the star. Here, b_V, b_B and b_U are the brightness of the star in the corresponding filters. Typically, for a blue star $(B - V) = -0.4$, for a red star $(B - V) = +2.0$ and for the sun $(B - V) = +0.53$.

Graph of surface temperature of stars against their brightness ratio or color index is shown below in Figure 1.4.

Once the brightness ratio of a star is measured using the filters, its surface temperature can be read from the graph. For the sun, $(B - V)$ is +0.53 and the corresponding brightness ratio, i.e. $\frac{b_v}{b_b} = 1.77$. This corresponds to a surface temperature of about 6000 K.

Surface Temperature of a Star

Surface temperature, also known as effective temperature or color temperature of a star, is the equivalent temperature of a perfect blackbody radiating at the same rate as the star. The star and the blackbody will have the same energy spectrum. According to Wien's displacement law, the effective temperature T_e of a star is inversely proportional to the wavelength (λ_m) at which its energy peaks.

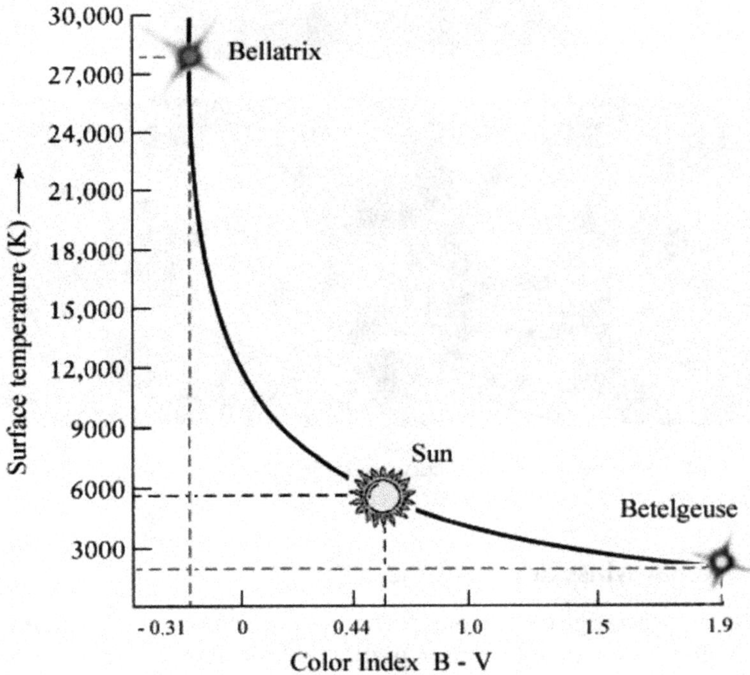

FIGURE 1.4 Surface temperature versus color index of stars

$$T_e \propto 1/\lambda_m, \text{ or } T_e \lambda_m = b, \text{ Wien's constant} = 2.89 \times 10^{-3} \text{ mK}$$

From the graphs in Figure 1.5, it is clear that for cooler stars the wavelength peak moves to higher wavelengths of the electromagnetic spectrum. The white vertical streak in the graphs corresponds to the optical range of the spectrum. Blackbody radiation from the Sun peaks at a wavelength of around 500 nm and its surface temperature works out to around 5780 K.

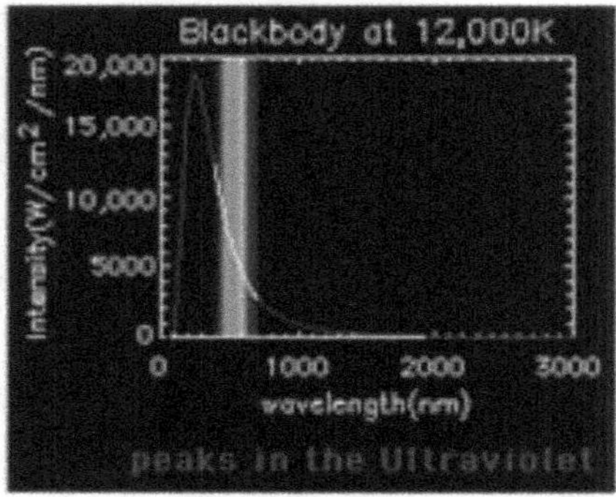

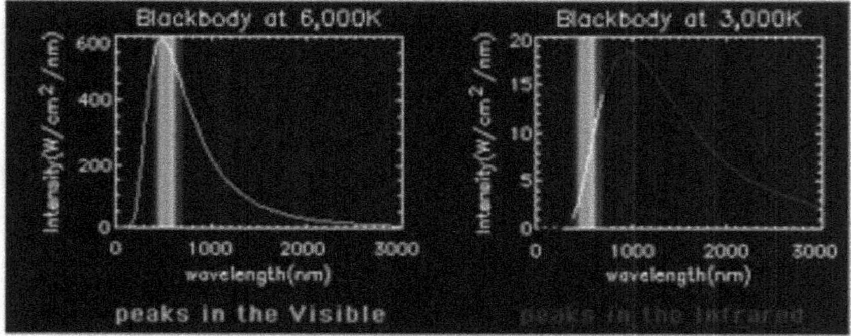

FIGURE 1.5 Energy distribution curve of a blackbody at different temperatures (Image Credit: NASA)

1.8 Stellar Mass and Lifetime

Masses of stars range from around 0.1 to 60 solar masses. If a gravitationally contracting cloud of interstellar matter is less than 0.1 solar mass, the core temperature does not reach high enough to initiate and sustain thermonuclear fusion reactions. On the other hand, in a star more massive than 60 solar masses, gravitational contraction is so rapid and so violent that the energy generated at the core tears the cloud apart much before a stable star is formed.

Mass of a star plays a key role in deciding its lifetime. A more massive star is a larger storehouse of energy, but has to convert its mass into energy at a faster rate to counterbalance the enormous gravitational force and therefore has a shorter life span. A less massive star burns its fuel at a much slower rate and hence lives longer. The Sun is half way through its lifetime of about 10 billion years. Sun is expected to live roughly for another 5 billion years before it extinguishes itself!

Binary Stars

Majority of stars in the sky are not single entities but consist of two stars revolving around their common center of mass, also known as barycenter. Such stars are said to form a binary system. The stars in a binary system are gravitationally bound to each other. Binary components are not necessarily of the same mass range. The more massive component controls the movement of the smaller star and its rate of revolution. By determining the rate of revolution of the smaller star, it is possible to estimate the mass of the other star. In 1844, German astronomer Friedrich Wilhelm Bessel found that Sirius, the brightest star, had a wobbly path and he concluded it should be because of the gravitational tug by a companion star. This companion, named Sirius *B*, was discovered by Alvan Graham Clark in 1862. Now it is established that Sirius *B* is a white dwarf.

Double stars are classified as visual binaries and spectroscopic binaries. Visual binaries are closer to the Earth and can be easily resolved into two separate point light sources. They are generally of low intensity and are of comparable mass. More than 70,000 visual binaries have been cataloged. For example, Sirius *A*, a main sequence star of 2 solar masses and Sirius *B*, a white dwarf of nearly 1 solar mass form a visual binary. Periods of revolution and size of each visual binary component can be determined through direct observation. Mass of a star can be arrived at if it is a member of a binary system. If M_1 and M_2 are the masses of the two component stars with a distance a between them and if P is their common period of rotation, then from Kepler's third law;

$$M_1 + M_2 = 2\pi a^3/GP^3$$

From this, the total mass of the stars can be estimated. If a_1 and a_2 are the distances of the stars from their common centre of mass, then

$$M_1 a_1 = M_2 a_2$$

Since $a_1 + a_2 = a$ mass of each individual component star can be calculated. Spectroscopic binaries are too far away from the Earth to be resolved. They are generally of high luminosity. They are identified as binaries from their spectra which reveal periodic Doppler shifts, both towards the red and the blue end, as the stars revolve around their common center of mass. The magnitude and direction of the Doppler shift helps in determining the speed and direction of motion of stars along the line of sight. With this, the period and radius of stellar orbits can be estimated.

A star is said to be born when it first starts emitting energy at least partly in the visible range and it can be seen as a point source of light in the skies. The star is said to be dead when it exhausts the fuel for nuclear burning at its core, loses its ability to radiate energy and cannot be spotted in the sky. Lifetime of a star is the duration for which it shines by emitting light of its own. Main sequence stars (refer section 1.9) not only have a wide range of luminosity but an even wider range of lifetime. A star's lifetime depends essentially on its mass at birth. Massive stars have larger mass and hence are more luminous since they have more energy available for conversion to radiant light. But this energy has to be expended at a faster rate in order to balance the enormous gravitational inward pull.

So while large mass M contributes in enhancing the lifetime of a star, its luminosity L tends to curtail it, i.e. the lifetime of a star $\propto \frac{M}{L}$. Mass-luminosity relation can be arrived at as $L \propto M^3$.

This leads to $\tau \propto \frac{1}{M^2}$. In other words, the lifetime of a star varies inversely as the square of its mass at birth. Large massive O-type stars (refer section 1.8) live for a million years, while the least massive M-type stars last for hundreds of billions of years!

1.9 Chemical Composition and Spectral Classification of Stars

Spectrum of star light consists of a continuous emission spectrum due to its hot central core. It is similar to that of a black body at the same temperature as that of the surface of the star. Radiation from the core passes through the cooler gaseous envelope, i.e. the chromosphere. Wavelengths specific to the elements present in the chromosphere are absorbed. This gives rise to a line absorption spectrum in the foreground of the continuous spectrum, similar to the Fraunhofer lines in the solar spectrum. The absorption lines are signatures of the elements present in the atmosphere of the star. The relative location and intensity of these absorption lines is an indicator of the chemical composition and relative abundance of the elements in the star's atmosphere, and serves as a yardstick for spectral classification of stars.

Photographic study of stellar spectra was initiated in 1885 by Edward Charles Pickering and his coworkers at Harvard University, under the guidance of Annie J. Cannon. In 1890, Pickering and his team classified the stars visible in the night sky into a continuous sequence based on their spectral pattern by the letters O, B, A, F, G, K and M (remembered with the mnemonic 'Oh! Be A Fine Girl, Kiss Me') from the hot blue stars to the cool red ones. Three subclasses of R, N and S (to continue with the mnemonic, 'Right Now Sweetheart') were later added to include stars cooler than *M*-type. (Refer Table 1.1 for details). *R* and *N*-type stars are now subsumed into the new *C*-type as *C-R* and *C-N* stars. Each class is further subdivided into 10 groups, called the Harvard sequence based on strength of H I lines. The subgroups are specified by the numbers 0 to 9 denoting a decrease in the H I line intensity. 99% of the stars are between the *B* and *M* spectral class (Table 1.1).

O stars: Spectral signature of these stars is the presence of strong lines of ionized helium. Ionization energy of He is maximum for any element at 25 eV and it indicates a high surface temperature of the star in the range of 30000 to 50000 K.

B stars: These have lower surface temperature which is inadequate to ionize helium atoms. Therefore, He II lines are absent in their spectra. Neutral He I lines are predominant along with ionized O II lines.

A stars: Neutral He I lines are absent. Strong H I Balmer lines and fainter lines of ionized metals are seen.

F stars: Lines due to neutral metals appear for the first time. Strong lines of ionized Ca II are seen at the blue end of the spectrum. Strong lines of neutral iron and other metals are also seen.

G stars: Strong absorption lines due to elements are seen, Ca II lines being the strongest.

K stars: Densely packed lines due to neutral metals are observed.

M stars: Absorption bands due to elements still in a molecular state, because of low surface temperature of the stars, are seen.

The spectral type of a star is completely described using Yerkes Luminosity classification as follows:

Ia: Bright supergiant **Ib: Fainter supergiant**

II: Bright giant **III: Normal giant** **IV: Sub giant**

V: Main sequence star **VI: White dwarf**

A star is completely designated by its spectral class followed by Harvard sequence number and spectral type.

Some examples:	Sun	G 2 V
	Rigel	B 8 Ia
	Vega	A 0 V
	Polaris	F 8 Ib
	Arcturus	K 2 III

TABLE 1.1 Spectral type classification

Spectral type	Surface temperature range (K)	Prominent spectral lines	Color of star	Example
O	30,000-50,000	Strong lines of He II	Blue	Zeta Puppis
B	10,000-30,000	He II lines are absent. Prominent He I and O II lines	Bluish white	Rigel
A	8,000-10,000	He I lines are absent. Strong H II Balmer lines along with fainter lines due to ionized metals	White	Sirius
F	6,000-8,000	Strong line of Ca II at the violent end and lines of neutral metal atoms	Yellowish white	Polaris
G	4,500-6,000	Strong lines due to elements, Ca II being the strongest	Yellow	Sun
K	3,500-4,500	Densely packed lines due to neutral metals	Orange	Arcturus
M	2,000-3,500	Molecular absorption bands	Orangish red	Barnard's star

1.10 Hertzspurng-Russell (HR) Diagram

Stars come in a large variety of temperature, luminosity, size and mass. In order to arrange these stars in an orderly manner, Ejnar Hertzsprung, a Danish astronomer and Henry Norris Russell, an American scientist independently proposed in 1910 that the absolute brightness of a star measured in terms of its absolute magnitude should be closely related to its color, which is indicative of its surface temperature. In 1913, they independently plotted the absolute magnitude of visible stars in the sky against their effective temperature. The resulting graph is referred to as Hertzsprung-Russell diagram or HR diagram or HRD. It is also known as color-magnitude diagram or CMD. HRD establishes the relationship between absolute magnitude, luminosity, spectral classification and surface temperature of stars. Also, the evolution of a star can be traced in HRD. It is used as a tool by astronomers to roughly estimate the distance of star clusters.

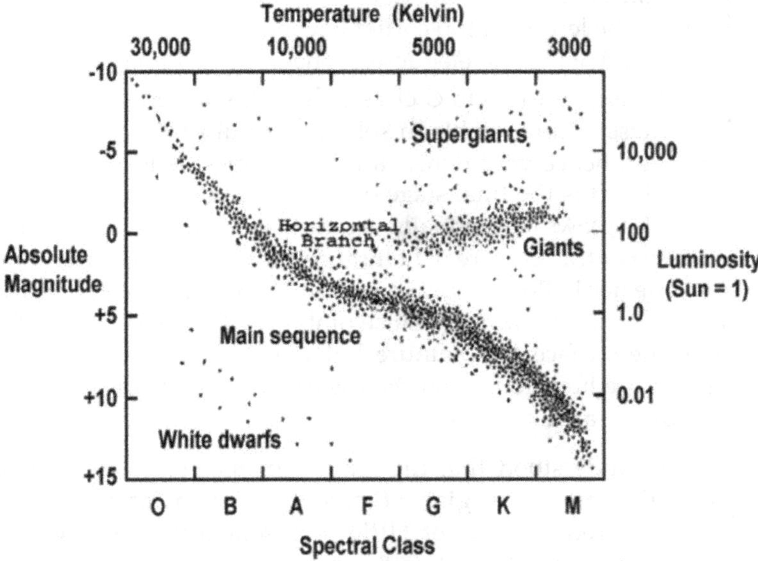

FIGURE 1.6 HR diagram (Credits: NASA)

It is seen from Figure 1.6 that 90% of the stars lie along a narrow diagonal band running from the top left to the lower right corner in the HRD. This band is called the Main Sequence Band. It is so called since the stars spend most of their lifetime at some location of this band. This band is also known as the Zero Age Band since the nursery of all newborn stars can be found along this band. A star is said to be born when it starts shining by emitting light of its own. This star will approach the main sequence band from the right. Heavier stars contract more rapidly and approach the main sequence band faster than the low mass stars. The sun, which is around 5 billion years old, can be located in the middle of the main sequence band.

There are sparsely spaced clusters of stars in different stages of their evolution in specific regions in HRD:

1. *Sub Giants* occupy a branch above the main sequence band between the *F* and *K* class of stars. These are stars which are brighter than main sequence stars of the same spectral class.
2. *Giant stars* are found above the sub giant branch between *G* and *M* stars. These are more luminous and larger than main sequence stars of the same surface temperature. They are also much brighter than the sub giants.
3. *Supergiants,* which are exceptionally luminous and very massive, are found in the left hand top corner beyond the range of *O* to *F* class of stars. E.g.: Antares, Beteleguese, alpha Scorpil, alpha Cygni.
4. *White Dwarfs* in the *B* to *G* class occupy the lower left hand corner. Their mass is comparable to solar mass but they are of the size of Earth and hence very dense, a million times denser than the Sun. White dwarf is the final stage of a dead medium-mass star like the Sun and it glows by gradually exhausting the fuel at its core. It is the severely contracted core of helium or carbon left over after the star's envelope has ballooned out. It is called 'white' because of its whitish appearance and 'dwarf' because of its size. Giants and Dwarfs have the same surface temperature but the giants are nearly a billion times more luminous than the dwarfs because of their larger radii and surface area.

It is now understood that the main sequence stars, on exhausting their fuel at the core, die as giants first and later as dwarfs in the cosmic graveyard. Different regions of HRD represent different stages in the evolution of a star from its birth to death.

Main sequence stars, also known as normal stars, have the following general characteristics:

1. All newborn stars can be located on the main sequence band of HRD.
2. All the live stars spend most of their lifetime in the main sequence band.
3. Temperature and luminosity of the star is related through Stefan-Boltzmann law.
4. Normal stars differ from each other only with respect to their mass. The reddest dimmest stars have a mass of around 0.1 solar mass, while the bluest brightest stars have a mass of around 60 solar masses.
5. Mass is the only criterion that decides the luminosity and surface temperature of a normal star.

6. The *p-p* cycle is the source of energy in the lower main sequence stars, while those stars in the upper main sequence are powered by the CN cycle.

1.11 Stellar Evolution

Let us briefly go through the various stages in the life of a star and trace it on the HR diagram. Enormous distended clouds of hydrogen along with dust particles are present in different parts of the interstellar medium in a galaxy. Swaying shock waves with a gigantic span arising out of cosmic explosions like the supernovae accidentally initiate the condensation of a gas cloud, also called a nebula. This cloud that is held together by its own gravity is still not visible and is known as a protostar. It is estimated that the cloud should consist of at least around 10^{57} particles weighing around 10^{30} kg to form an average mass star like the Sun. If gravity were the only force acting on a protostar, it would theoretically get crushed to a point. There must be a radially outward force to counter gravity since the contracting cloud does not diffuse with time. Pressure associated with the gas molecules is insufficient to balance the inward gravitational force. Over millions of years, as the cloud keeps contracting, material at the core of the cloud heats up and it gets more and more dense. Eventually, the core temperature becomes high enough to initiate thermonuclear fusion of hydrogen to helium. This ignition sends radially out a burst of high energy radiation. This has the effect of ballooning out and cooling of the protostar. This halts the fusion reactions. Protostar material again starts contracting till the fusion cycle starts all over again. After several cycles of alternate contraction and expansion over several thousand years, the core temperature reaches around 15 million kelvin and hydrogen fusion becomes a self sustaining reaction. The protostar appears as a self luminous celestial object, and we say, "A star is born!" The time scale from the protostar to the normal star depends on the initial mass of the contracting cloud. A star heavier than the Sun has a shorter time scale. Once born, a star's life is a continual battle to achieve a balance between gravitational contraction and radiation pressure!

Once born, the star attempts to attain a critical state of hydrostatic equilibrium between its outward radiation pressure and inward gravitational pull. This forces the protostar through a phase of irregular changes in its luminosity. This stage of evolution is referred to as the *T* tauri variable star. Once perfect hydrostatic equilibrium is established, the rate of decrease of temperature with respect to radial distance becomes constant. In the Sun for example, this temperature gradient is about 0.021

kelvin per meter. Once thermal equilibrium is reached between energy generated at the core due to fusion of hydrogen through *p-p* and CNO cycles and the total energy radiated from the surface of the star, the star is identified as a 'normal star' or a 'main sequence star'. What happens to a star once the fusion reactions stop at its core depends on the initial mass of the star.

Once hydrogen in the core depletes below a critical fraction, the star undergoes what is known as virial contraction causing the core to shrink, and the core temperature and pressure to rise. The hot core ignites the hydrogen shell of the star. The burning envelope keeps expanding slowly and the star veers off upwards to the right of the main sequence band in the HR diagram for the first time. With mounting temperature at the core, the shell loses its ability to absorb all the energy pouring out, becoming transparent and allowing most of the radiant energy to reach the outer layers of the star. This raises the luminosity of the star by a factor of several hundred, while its temperature would have fallen to a mere 3000-4000 kelvin. This bloated star now looks like a giant, glowing with a typical red color. This stage in the evolution of a star is known as the Red Giant.

When the contracting helium core reaches a temperature of around 100 million kelvin with a corresponding density of 10^8 kgm^{-3}, helium nuclei start fusing into carbon nuclei through the triple alpha or 3α process. The core of burning helium is surrounded by a shell of burning hydrogen and outer shell of non-burning hydrogen. The energy released is large enough to engulf the entire core. As a result, the core explodes aggressively within a few seconds. This event in the life of a star is referred to as Helium flash! With this, the star cools along with its shell. Its radius decreases and its luminosity drops. The plot of the star in the HR diagram moves back towards the main sequence band.

Helium fusion results in a carbon core, which in turn culminates in a carbon flash, and so on. The onset of each stage of nuclear burning at the core with a new fuel sends the star zooming more rapidly towards the Red Giant stage. Each subsequent stage, however, is characterized by a lower energy input and therefore lasts for a shorter interval of time. This cycle continues till iron is formed at the core. Fusion of iron to a heavier element is an endothermic reaction which the star does not favor. Not all stars go through all the fusion cycles till iron is formed at the core. Heavier stars sustain more fusion cycles.

Generally, stars with mass comparable to that of the Sun do not go beyond the helium burning stage. Temperature of the contracting core rises to a whopping 10^8 K with a corresponding density of 10^9 kgm^{-3}. Heat generated during contraction causes the star's envelope to balloon out gradually and cool. The severely contracted core of helium is called a White dwarf, white because of its appearance and dwarf because of its size. The star's plot in the HR diagram, called the Hayashi track, crosses the main sequence band and reaches the bottom left hand corner. A white dwarf typically radiates in the UV and illuminates the expanding shell surrounding it. This brilliantly colored shell is known as Planetary Nebula. With ageing, a white dwarf cools by radiating its thermal energy to yellow, red and eventually black! Indian-American astrophysicist Subramanyan Chandrasekhar worked out the upper mass limit for a white dwarf as 1.44 solar mass, now famous as Chandrasekhar mass limit, by invoking Einstein's special theory of relativity and quantum principles. It can be shown that the radius R of a white dwarf is related to its mass M using the relation $RM^{1/3}$ is constant. Interestingly, it predicts that more massive white dwarfs are smaller in size! A spoon of white dwarf material would weigh close to 15 tons!

Core temperature of a more massive star reaches around 10^9 kelvin at which nuclear burning of carbon is supported. This generates so much heat that the outer layers of the star explode in a cataclysmic event called Supernova explosion or merely Supernova or Exploding star. This event ejects stellar matter into space at enormous speeds of 7000 kms^{-1}. Supernova expands faster than a planetary nebula and is shorter lived. Core collapse during supernova explosion is so violent that the white dwarf stage is totally bypassed. Supernova will shine as brilliantly as a whole galaxy if the debris of the explosion contains matter equivalent to just one solar mass! Elements heavier than iron form during supernova explosion.

Collapsed core of a supernova explosion will end up as either a neutron star or a blackhole depending on the mass of its contents-3 solar masses being the upper limit of a neutron star, while that for a blackhole is around 50 solar masses! Figure 1.7 pictorially depicts the various stages in the evolution of low-mass and high-mass stars.

Generally, stars with an initial mass of around 8 solar mass end up as neutron stars. Electrons inside a neutron star combine with protons through inverse beta decay $p + e^{-1} \rightarrow n + v_e$ to form neutrons (v_e being electron-neutrino) and hence the name. Quantum mechanical considerations shows that the relation $RM^{1/3}$ is constant. Density of a neutron star is comparable to nuclear density $\simeq 10^{18}$ kgm^{-3}! A teaspoon of neutron star would weigh a billion tons!! Neutron stars were ignored for over 30 years after their discovery. In 1967, Jocelyn Bell and Anthony Hewish discovered an exotic pulsar emitting regular pulses of radio waves every 1.377 seconds. This puzzled the scientific community till it was concluded that pulsars are nothing but rotating neutron stars pulsing radio waves similar to a beacon of light from a lighthouse.

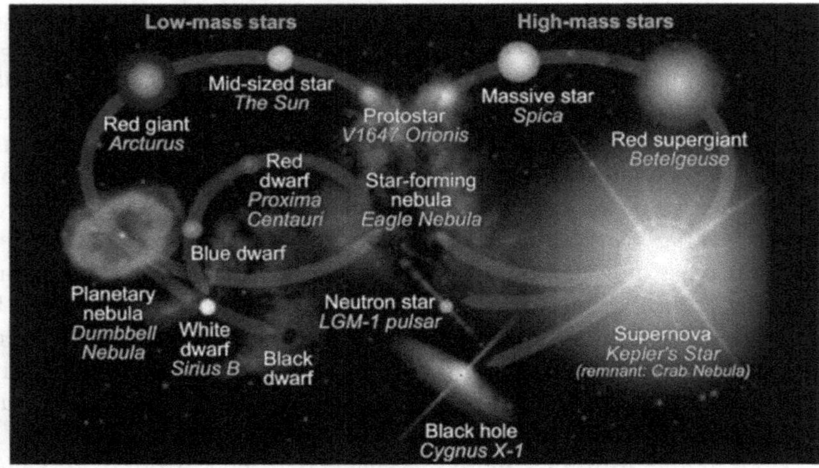

FIGURE 1.7 Life cycle of the stars (Credits: cmglee/NASA Goddard Space Flight Center)

1.12 Atmospheric Windows

The Sun radiates energy that spans the entire electromagnetic spectrum. Solar radiation keeps bombarding the Earth constantly throughout the day and night. Solar radiation has to pass through the various layers of the atmosphere before reaching the Earth. The radiation gets either transmitted, reflected or absorbed by the gases and water vapor present in the various layers of the Earth's atmosphere. The extent of absorption depends on the physical and chemical composition of the layer. IR rays are absorbed mainly by water vapor and carbon dioxide present within 30 km above the Earth. Ozone found in layers beyond 35 km strongly absorbs UV radiation from the Sun and helps in shielding life on Earth from its harmful effects. Radiation with wavelengths shorter than 300 nm or energy greater than 5 eV are effectively blocked by molecular nitrogen, atomic oxygen and nitrogen, water vapor, ozone and carbon dioxide in the atmosphere, thus protecting us from exposure to a range of highly energetic and ionizing solar radiation. Exposure to high energy radiations can damage the cells in organic matter. These radiations also have curative properties when used in controlled doses, for example, to kill cancer cells. The atmospheric blanket helps in protecting life on Earth and in maintaining the planet habitable.

The wavelength ranges of solar radiation for which the Earth's atmosphere is particularly transmissive, i.e. transparent, are referred to as atmospheric windows. The first window is in the wavelength range 400 nm to 720 nm and is called the optical window (refer Figure 1.8). This makes optical astronomical observation the most preferred technique for analyzing starlight. What we know of the universe today is built up from mostly the light received from the stars and other celestial objects (Ronan 1964). But the optical window is a very narrow portion of the electromagnetic spectrum. Surveys of the sky at different wavelengths have revealed many objects in the universe that emit very little in optical wavelengths but are very bright in infrared or X-ray wavelengths. Valuable information about sky objects, stars in particular, is hidden in the various spectral regions and hence observing them in different wavelength zones of the spectrum is essential. The second window extending from 1 cm to 100 m is the radio window. Radio astronomy developed from 1932 onwards.

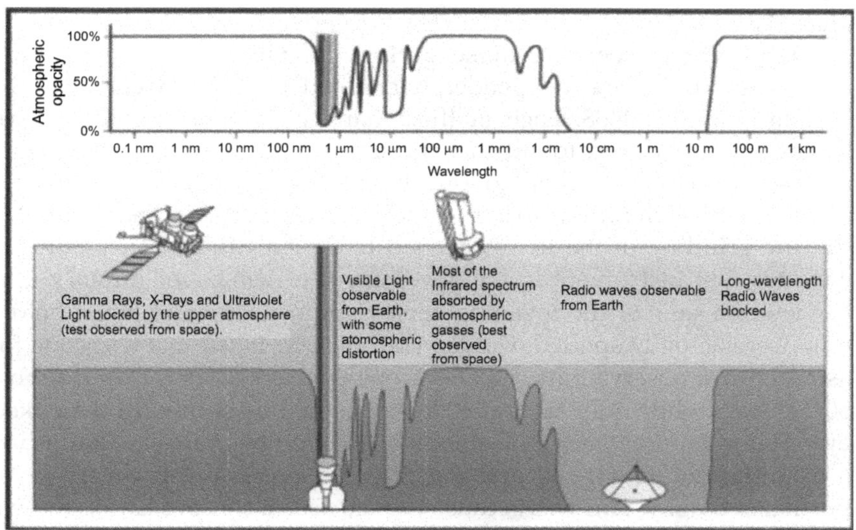

FIGURE 1.8 Atmospheric transparency to insolation (Credits: NASA, SVG by Mysid)

The atmosphere is transparent to most infrared radiation, at least partially. Atmospheric transparency is good in IR between 0.8 µm to around 40 µm. The UV spectrum of a star is particularly useful in revealing the abundance of elements in its atmosphere. Thus, in order to gain as much information as possible about a star, telescopes have to be taken into space above the Earth's atmospheric curtains, beyond 150 km, where the atmosphere is transparent to all wavelengths of the spectrum. This is the reason why UV, X-ray and gamma ray astronomy became possible only over the last 35 years after the advent of space vehicles and rockets capable of transporting telescopes, spectrographs and other observational devices into space.

1.13 Optical Telescopes

Human eyes with an average size of its pupil as 5 mm can only view stars upto 6th magnitude. Telescope is the most widely used instrument for astronomical observations of sky objects fainter than 6th magnitude. Optical telescopes are used for observations in the optical range of 400-800 nm. Galileo's simple telescope with an aperture of 2.5 cm could capture fainter stars upto 9th magnitude. Since then, telescopes have been designed with progressively larger apertures to gather as many photons as possible from the sky object being studied. The 5 m telescope in Mount Palomar Observatory, USA can detect stars right upto 25th magnitude, 10^8 times fainter than a 6th magnitude star. Light from the telescope can either be viewed with a digital camera or recorded on a film for a detailed analysis.

Three Dutch spectacle makers: Hans Lippershey, Zacharia Jansen and James Metius are independently credited with the invention of the first telescope in 1608. Each of them happened to view through two lenses held apart and found that the image was more magnified than through a single lens. This aroused the interest of an Italian astronomer Galileo Galilei who was also a professor of Mathematics at Padua. Galileo did the grinding of the lenses needed to design his first telescope. He gave his first instrument to the government of Venice for military use- to view the sails of the invading enemy ships a couple of hours before they were actually spotted. Galileo was the first to use the telescope for astronomical observations. The telescope enabled Galileo to see almost 50,000 stars while only around 5000 stars could be seen with the naked eye. The main drawback of Galileo's telescope however was that it had a magnification of only 30, and also, it could magnify only a very small portion of the sky. This was because Galileo used a concave lens as the eye lens which refracted the rays outwards. Only a portion of the spread out rays formed the final image which ended up being blurry and distorted. But this did not deter Galileo's enthusiasm from keenly observing the craters, mountains and valleys on the moon, Jupiter with its four moons, sunspots and phases of Venus for the first time. Invention of the telescope revolutionized observational astronomy.

Apparent Size and Angular Magnitude of an Object

An object which is far away appears to be smaller in size. It seems to grow in size as we walk towards it, say a tree on the roadside. Apparent size of an object depends on its angular magnitude-the angle subtended at the eye by the cone of rays originating from it. So magnification of an object can be achieved either by producing its enlarged image or by bringing it closer to the eye.

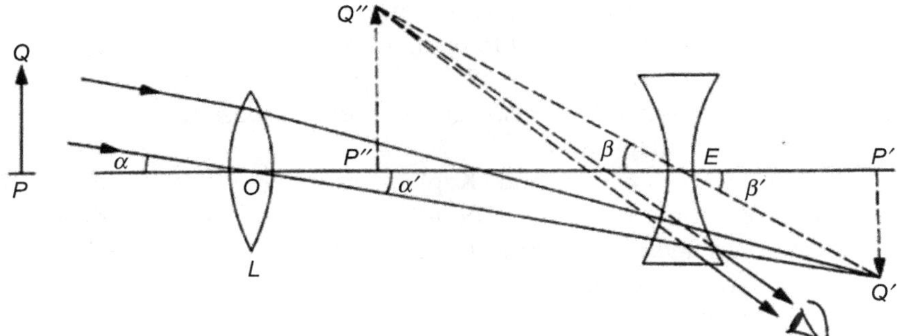

FIGURE 1.9 Galileo's refracting telescope (Credit: Verma)

In the ray diagram shown in Figure 1.9 for Galileo's refracting telescope, *PQ* is the object and the convex lens *L* facing the object is the object-glass or objective. Parallel rays of light from the distant object *PQ* passing through the objective converge to form a real and inverted image at *P′Q′* in the focal plane of *L*. But these rays are intercepted by the concave eye lens. The rays striking the eye lens are rendered divergent and appear to be coming from *P″Q″*, which is the final, erect image of *PQ*. Since the incident rays making an angle α with the principal axis emerge at a greater angle β, the apparent angular size of the final image increases and is responsible for the perceived magnification. Maximum magnification produced by a Galilean telescope is 30.

Johannes Kepler, who was in constant correspondance with Galileo, came up with another design of telescope. He replaced the concave lens of the eyepiece with a convex lens. All the light refracted through the objective could now be captured by the eye lens to provide a larger and clearer field of view except that the final image was inverted. This does not pose a problem to astronomers since they all look at the Sun, moon and planets which are spherical, and stars that are point sources of light.

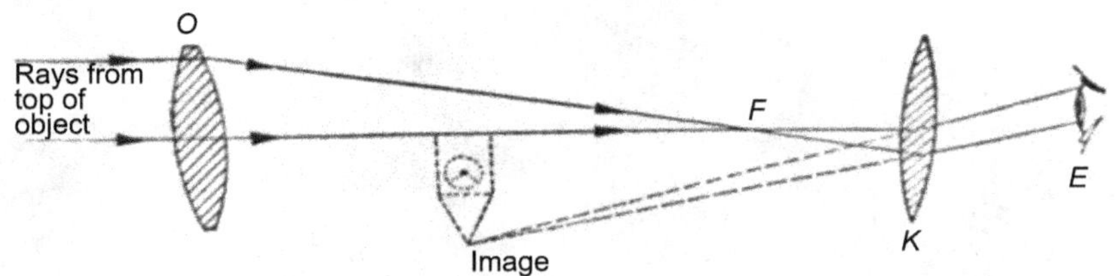

FIGURE 1.10 Kepler's refracting telescope (Credit: David Wheeler)

Kepler's telescope (Figure 1.10) was widely adopted by the astronomical community. The only disadvantage of the telescope was that it produced images that had colored edges. This was because a single lens was used as the object-glass which failed to bring to a common focus all the colors present in the incident light. In 1757, an optician by name John Dolland overcame this problem by constructing an object-glass using two lenses made of different materials.

Telescopes of Galileo and Kepler are referred to as refracting telescopes or simply refractors. Though versatile in operation, they have some disadvantages: as the aperture size increases, the device becomes more expensive since the cost of production of high-quality lenses goes up. Also, the tube length increases and the telescope as a whole becomes heavy, posing safe keeping and transportation problems. It is technically difficult to manufacture huge lenses with perfect curvature on the outside and no imperfections on the inside. Since the lenses are supported only around their rims, they tend to sag under their own weight. All images produced with a refracting telescope suffer from chromatic aberration-the rainbow of colors round the image. This is inherent with all refracting lenses and cannot be totally compensated for.

Sir Isaac Newton, who had done some extensive work in optics, suggested a novel but simple way of overcoming the problems with refractors. He suggested the use of a tiny mirror to collect the incoming light from a star and reflect it on to the eyepiece, as shown in Figure 1.11. This gave birth to the first reflecting telescope whose images were free from chromatic aberration and came to be acclaimed as more superior to refracting telescopes.

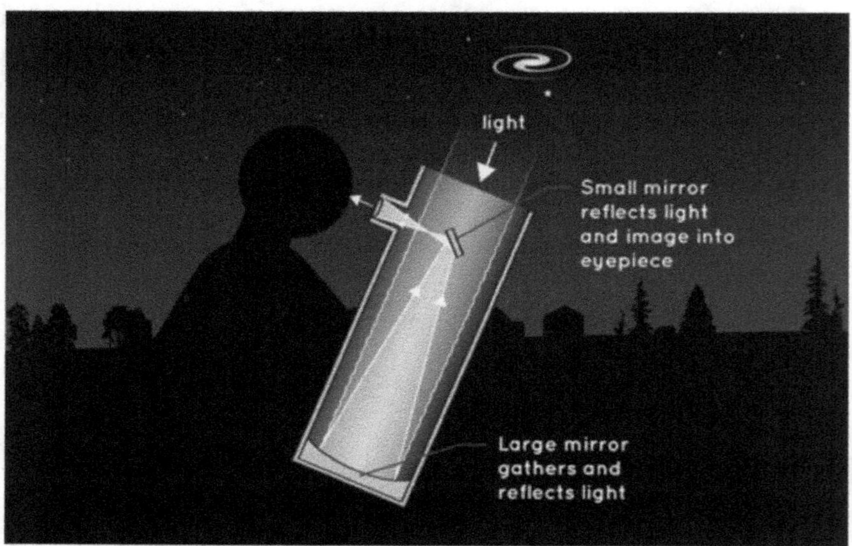

FIGURE 1.11 Newton's reflecting telescope (Credits: NASA)

In 1774, William Herschel began designing reflecting telescopes and used them to watch the sky. He gave up his successful profession in music in favor of astronomy. In 1781, Herschel created a sensation when he chanced upon seeing a new planet orbiting the sun. This hitherto unseen planet was called Uranus. Mankind, which was familiar with only five planets visible to the unaided eye-Mercury, Venus, Mars, Jupiter and Saturn, stood in awe of Uranus. Herschel kept making larger and larger telescopes, the largest being 40 feet long with a mirror almost 4 feet in diameter and weighing nearly a ton. A very vast field of view is possible with large telescopes, though the objects look dimmer.

All telescopes have the following common attributes which define their operation and performance:

***Aperture*:** It is the diameter of the component of the telescope, objective lens of a refracting telescope or the primary mirror of a reflecting telescope, through which light from an object is gathered. Larger aperture results in a brighter image. Fainter objects can be seen with telescopes of larger aperture. Thus, aperture decides the light gathering power of the telescope and hence defines the faintest sky object that can be seen through it. Mounting and transportation is, however, difficult with telescopes of large apertures.

Light gathering power: The objective is the main light gathering component of a telescope. Light gathering power of a telescope is directly proportional to the area of the mirror or the lens in its reflector or refractor. For context, the maximum aperture of a normal human eye is around 7 mm. So a moderate telescope with 100 mm aperture will increase the light gathering power of the eye by $(100/7)^2$, i.e. 204 times.

Magnifying power: In order to cast an image comfortable for viewing, every telescope is fitted with an eyepiece which is either a single lens or a combination of lenses. It is through the eyepiece that we look at an object. Magnifying power of a telescope is decided by the focal lengths of the lenses used as objective and eyepiece. It is given by the ratio of focal length of the objective to that of the eyepiece. For example, a telescope with a 12″ focal length objective and a 2″ eyepiece produces a magnification of (12″/2″) = 6*x*. Most telescopes allow the replacement of its eyepiece to get the desired magnification (Ventrudo 2016). However, the maximum useful magnification of a telescope before the image becomes dim and fuzzy is around 50 times its aperture in inches.

Resolving power: This defines the ability of a telescope to see as separate two objects which are close to each other. In other words, resolving power decides the finer details of the object being viewed. Resolving power is defined as the least angular separation between two stars at which they are just distinguishable. Resolving power of a telescope is inversely proportional to aperture of the lens or the mirror used in it. Resolving power of a telescope with an aperture of D expressed in millimeters is given by ($116/D$) arcseconds. Theoretical resolving power of a 1 m aperture telescope is 0.1 minute of an arc.

Focal ratio or f-ratio: It is the expedient of dividing the focal length of the reflector or refractor of a telescope by its aperture (Wall 2018). Focal ratio is designated by *f*/number. For example, for a refractor with an aperture of 10 cm and focal length of 50 cm, focal ratio is 50/10 = 5 and the telescope is said to have an *f*/number of *f*/5. Focal ratio gives the relative 'speed' of the telescope. A higher focal ratio implies higher magnification and narrower field of view. A focal ratio of 10 is ideal for viewing the moon, solar planets and binary stars. However, it is advisable to use a telescope of *f*-ratio 7 or less for a wider field of view required for viewing the Milky Way, other galaxies and star clusters.

Worked Examples

1. The star ε Eridani is at a distance of 3.27 pc from the Earth. Calculate its luminosity if its apparent brightness is 9.454×10^{-10} Wm^{-2}.

   Solution: Distance of the star:

   $$d = 3.27 \text{ pc} = 10.14 \times 10^{16} \text{ m}$$

   If L is luminosity of the star, its apparent brightness:

   $$b = \frac{L}{4\pi d^2}$$

   $$\therefore L = 4\pi d^2 b = 4\pi(10.14 \times 10^{16})^2 (9.454 \times 10^{-10}) = 1.221 \times 10^{26} \text{ W}$$

2. The star Sirius A has an effective surface temperature of 9500 K and a luminosity of about 1.02×10^{28} W. Calculate the radius of Sirius star. Given $\sigma = 5.67 \times 10^{-8}$ Wm^{-2} K^{-4}.

   Solution: Using the luminosity relation, radius of the star is

   $$R = \sqrt{\frac{L}{4\pi\sigma T_e^{\,4}}} = \sqrt{\frac{1.02 \times 10^{28}}{4\pi \times 5.67 \times 10^{-8} \times (9500)^4}}$$

   $$R = 1.3 \times 10^{9} \text{ m}$$

3. Apparent magnitudes of the stars Wolf 359 and Cygni *A* are +13.45 and +7.49, respectively. Calculate their relative brightness.

Solution: Apparent magnitudes of Wolf and Cygni, respectively, are given by the equations $m_W = C - 2.5\ \log b_W$ and $m_C = C - 2.5\ \log b_C$ where C is a constant. Subtracting, $m_W - m_C = 2.5 \log\left(\frac{b_C}{b_W}\right)$

$$\log\left(\frac{b_C}{b_W}\right) = \frac{m_W - m_C}{2.5} = \frac{5.96}{2.5} = 2.384$$

$$= \frac{b_C}{b_W} = 10^{2.384} = 242.1$$

Cygni appears around 240 times brighter than the Wolf star.

4. The star Fomalhaut has apparent and absolute magnitudes of +1.17 and +1.74, respectively. At what distance from the Earth is the star located?

Solution: From the distance-modulus relation $(m - M) = 5 \log d - 5$

$$\log d = \frac{m - M + 5}{5} = \frac{1.17 - 1.74 + 5}{5} = 0.8860$$

Distance of Fomalhaut = $d = 10^{0.8860} = 7.6913$ pc = 7.6913 × 3.26 ly = 25.07 ly

5. Parallax of the star Proxima Centauri is 0.772 arc-sec. How long does light from Proxima Centauri take to reach the Earth?

Solution: Distance of the star = $d = \frac{1}{0.772} = 1.2953$ pc = 1.2953 × 3.26 ly = 25.07 ly

Light from Proxima Centauri takes around 25 years to reach the Earth.

6. The star Tau Ceti is at a distance of 11.90 ly. Calculate its parallax.

Solution: $d = 11.90 \text{ ly} = \frac{11.90}{3.26} \text{ pc} = 3.65 \text{ pc}$

Parallax of Tau Ceti = $p = \frac{1}{d} = \frac{1}{3.65} = 0.274$ arc-sec

7. Sun has a surface temperature of around 6000 K and a radius of 7×10^8 m. Calculate (i) the rate at which sun radiates energy, (ii) the number of 100 W bulbs match the luminosity of the Sun and (iii) Is it possible for a company to manufacture the required number of bulbs if its manufacturing capability is 100 million such bulbs in a year? Justify your answer.

   Given: Stefan constant = 5.67×10^{-8} Wm^{-2} K^{-4}

   Solution:

   (i) From luminosity relation, luminosity of the Sun is

   $$L = 4\pi\sigma T^4 R^2 = 4\pi \times 5.67 \times 10^{-8} \times 6000^4 \times (7 \times 10^8)^2 = 4.49 \times 10^{26}\ \text{W}$$

   (ii) Number of 100 W bulbs required to match solar luminosity = N

   $$= \frac{4.49 \times 10^{26}}{100} = 4.49 \times 10^{24}$$

   (iii) Time required to manufacture these bulbs = $T = \dfrac{4.49 \times 10^{24}}{100 \times 10^6} =$ 4.49×10^{16} years.

   The company will not be able to manufacture the required number of bulbs since the life of the sun and hence of everything on the earth is only around 5 billion years, i.e. 5×10^9 years.

8. Estimate the temperature of a star whose radiation peaks at 6563 Å, given Wien constant = 3×10^{-3} mK.

   Solution: According to Wien's displacement law,

   $$\text{Wien's constant } b = T_e \lambda_m \text{ or } T_e = \frac{b}{\lambda_m} = \frac{3 \times 10^{-3}\ \text{mK}}{6563 \times 10^{-10}\ \text{m}} = 4571.1\ \text{K}$$

9. Magnitude difference between the Sun and the full moon is 14. What is their brightness ratio?

   Solution: Magnitude difference of 1 corresponds to a brightness ratio of 2.512.

   Brightness ratio between the Sun and the full moon is 2.512^{14} = 3,98,107.

   In other words, the Sun is around as bright as 4 lakh full moons!

10. A star has absolute and apparent magnitudes of 4 and 17, respectively. How long does it take for light from the star to reach us?

Solution:

(i) First method

Substituting $m = 17$ and $M = 4$ in the distance-modulus equation $m - M = 5 \log d - 5$ or $\log d = (m - M + 5)0.2$ we have,

$\log d = (17 - 4 + 5)\ 0.2 = 3.6$

$\therefore d = 10^{3.6}$ pc = 3961 pc = 3961 × 3.26 ly = 12,913 ly

Light from the star takes 12,931 years to reach the Earth.

(ii) Second method

Distance modulus of the star: $m - M = 17 - 4 = 13$

Break this into 5's and 1's: $13 = 5 + 5 + 1 + 1 + 1$

Now transform from a magnitude difference to a brightness ratio, i.e. for every 5 write 100, for every 1 write 2.5 and replace every + with the multiplication sign *x*.

$$\text{Brightness ratio: } \frac{B\,(\text{at 10 pc})}{B\,(\text{at distance } d)} = 100 \times 100 \times 2.5 \times 2.5 \times 2.5$$

$$= 156250$$

Distance of the star: $d = (10 \text{ pc})\sqrt{156250}$

$$d = 10 \text{ pc} \times 395.3 = 3953 \text{ pc} = 3953 \times 3.26 \text{ ly} = 12888 \text{ ly}$$

Light from the star takes 12,888 years to reach the Earth.

11. Absolute magnitude of a distant galaxy is –22 and that of the sun is +4.76. Estimate the number of stars in the galaxy assuming each star is as luminous as the Sun is.

Solution: Magnitude difference between the galaxy and the Sun is $4.76 - (-22) = 26.76$.

Brightness ratio between the galaxy and the sun is $2.512^{26.76} \simeq 5 \times 10^{10}$.

The galaxy has around 50 billion stars.

12. Determine the distance of Sigma Draconis if its parallax is 0.17 arcsec.

Solution: Distance of the star in pc is given by $d = \dfrac{1 \text{ AU}}{0.17} = 5.88$ pc

$$= 5.88 \times 3.26 \text{ ly} = 19.17 \text{ light years}$$

Exercise

1. Calculate the apparent brightness of the Sun, also known as solar constant, given the mean distance between Earth and the Sun is 1.5×10^{11} m, if solar luminosity is 3.9×10^{26} W.

   Answer: 1380 Wm^{-2}

2. Star Sirius *A* has a luminosity of 1.02×10^{28} W and apparent brightness of 1.22×10^{-7} Wm^{-2}. How far is it from the Earth?

   Answer: 81.4×10^{15} m

3. If the distance modulus of the star Altair is –1.44, calculate its distance from the Earth.

   Answer: 16.8 ly

4. Star Proxima Centauri has a parallax of 0.772 arc-sec. Calculate its distance from the Earth.

   Answer: 4.223 ly

5. A star with an apparent magnitude of +12 is located at a distance of 50 ly. What is its absolute magnitude? What is its luminosity in terms of solar luminosity? Given apparent magnitude of the Sun = –26.74.

   Answer: $M = 16.07$, $L = 7.5 \times 10^{-18} L_{\odot}$

6. Betelegeuse, a red star in the constellation of Orion, is 10^4 times as bright as the Sun and has a surface temperature of 3000 K. Calculate its radius in terms of solar radius, given effective temperature of Sun = 6000 K.

   Answer: $R = 400 R_{\odot}$

7. What is the absolute magnitude of a star whose apparent magnitude is +3.3 and parallax is 0.025 arc-sec?

   Answer: 0.290

8. Luminosity and surface temperature of the Sun are 4×10^{26} W and 5800 K, respectively. Calculate its radius, given that the Stefan constant = 5.67×10^{-8} Wm^{-2} K^{-4}.

   Answer: 7×10^{-8} m

9. What is the absolute magnitude of the Sun if its apparent magnitude is –26.33? Distance between the Sun and the Earth is 1.5×10^{11} m.

   Answer: +4.74

10. What is the magnitude difference between two stars whose luminosity ratio is 1589?

    Answer: 8

CHAPTER 2

Solar System Planets

'Our species needs and deserves a citizenry with minds wide awake and a basic understanding of how the world works.'

– Carl Sagan

2.1 Introduction

Man has always been curious to explore the dark, eerie and seemingly peaceful space, basically to get to know his cosmic neighbors and to establish a cosmic identity for himself. From time immemorial, all keen observers had noticed the rhythmic parade of seven celestial objects across the night sky, namely, the Sun, the moon and the planets Mercury, Venus, Mars, Jupiter and Saturn, all of which could be seen with the unaided eye. The Hebrews named each day of the week after these celestial bodies. The Greeks referred to them as *'planets'* meaning *'wanderers'* (Rana and Wilkinson 1989). Modern astronomers redefined the concept of planets to include all celestial objects that orbited the Sun in near circular orbits. In 1781, German-born British astronomer Sir William Herschel discovered the planet Uranus, while the German astronomer Johann Galle discovered Neptune in 1846. Today we know the solar system as consisting of the parent star, the Sun around which eight planets orbit. Planets in sequence of their increasing distance from the sun are remembered with the famous mnemonic, *"My (Mercury) Very (Venus) Eager (Earth) Mother (Mars) Just (Jupiter) Served (Saturn) Us (Uranus) Noodles (Neptune)"*. Mercury, Venus and Mars resemble the Earth in their composition and along with Earth are known as ***inner planets*** or ***terrestrial planets***. Jupiter, Saturn, Uranus and Neptune are 5 to 10 times larger and more massive than the Earth but are less dense because of their gaseous composition. Collectively, these four ***outer planets*** are known as ***Giant planets*** or ***Gas giants***.

2.2 Origin of Solar System

Several dozens of theories have been put forward by avid scientists to explain the origin of the solar system. Any theory is expected to account for the following salient features of the solar system (Rana and Wilkinson 1989):

1. The Sun accounts for 99.8% of the mass of the solar system.
2. Remaining 0.2% of the mass of the solar system, distributed among planets, comets and asteroids orbiting around the Sun, account for 98% of the total angular momentum of the solar system. This implies that the Sun rotates very slowly.
3. All planets go round the Sun in the same sense as the rotation of the Earth.
4. Orbits of all planets are confined to within 25 degrees of the equatorial plane of the Sun.
5. Inner and outer planets of the solar system have different compositions; while the inner planets lack volatile gases, hydrogen and helium – the two most abundant elements in the universe, two of the giant planets, Jupiter and Saturn, have chemical composition akin to the Sun itself.
6. Asteroids and meteorites have similar chemical composition.
7. The Moon and Mercury stand in evidence of heavy meteoritic bombardment.

As per the Nebular theory proposed in 1734 by the Swedish scientist Immanuel Kant, the solar system was born out of an enormous interstellar cloud of gas and dust particles. The word *'nebula'* in Latin means *'cloud'*. This solar nebula consisting of around 10^8 to 10^9 molecules per cubic meter kept contracting under its own gravity and over a period of time there was a rise in its internal temperature and pressure. Also, as the cloud became hotter due to contraction, it kept radiating energy. As the central cloud density increased to beyond 10^{16} molecules/m^3, the constituent particles started colliding with each other and began growing in size. This happened in the outer parts of the cloud as well. As the cloud kept spinning faster in order to conserve its angular momentum, the outer region gradually separated itself and formed a protoplanetary disc around the core, which initially had attained the surface gravity of the sun. The Sun kept accreting matter from the disc along spiralling streamers. Minor instabilities in the disc led to the formation of localized planetesimals, which were typically rocky in the terrestrial planet zone, while the outer zone planetesimals mainly consisted of a mixture of ice, water, solid carbon dioxide, ammonia and methane. The moon is also surmised to be formed due to accretion of rocky planetesimals. The giant planets formed out of icy planetesimals. This explained why the orbitals of the planets lie in the equatorial plane of the sun. This nebular theory was further refined by the French scholar Simon Laplace towards the end of the 18th century.

Those planetesimals that failed to form planets disintegrated into comets, meteors, meteoroids, interplanetary dust particles including the asteroid belt that exists between the orbits of Mars and Jupiter. During this phase of disintegration, the moon and Mercury were continuously bombarded by meteorites. The main drawback of Kant-Laplace nebular theory was its inability to account for the range of diameters of the solar planets.

In 1919, Sir James Jeans, a British scientist, proposed the Tidal theory of formation of the solar system. Sir Harol Jeffreys, another British scientist, modified the theory in 1929. As per the tidal theory, an 'intruding star' of a much larger size moved so close to the primitive sun that both tidally interacted with each other. Since the approaching edges of the two experienced stronger gravitational attraction towards each other as compared to the outer edges, each assumed the shape of a prolate spheroid, ejecting a trail of hot gases (Woolfson 1978). These gases cooled as they expanded to form planetesimals, which eventually coagulated into planets and their natural satellites.

2.3 Solar Planets

Solar system consists of the parent star Sun with directly or indirectly gravitationally bound objects orbiting it. Solar planets which directly go round the sun from the nearest to the farthest are:

Mercury, the Roman ***god of travel,*** is one third the size of Earth and has 1/10th the Earth's mass. Being closest to the Sun, it is very difficult to observe Mercury from the Earth. Proximity to the sun has also made Mercury lose the lighter and volatile elements from its exosphere. Mercury has no moons. It takes 59 earth days to spin once about its axis and 88 days to orbit once around the Sun. Mercury has a slightly higher density as compared to other terrestrial planets. Surface of Mercury has a cratered surface just like the moon as material from its surface is blasted off by solar wind and meteoroid strikes. Under most favorable conditions, Mercury can be seen for around 1½ hours before sunrise or after sunset. With a telescope, however, it can be seen in broad daylight. Interestingly, Mercury reveals moon-like phases: from a new moon during inferior conjunction when it passes between the Sun and the Earth, to a full moon during superior conjunction while the Sun is between the Earth and Mercury.

Venus, Roman ***goddess of love,*** is Earth's nearest neighbor. It is about the same size and mass as Earth and is aptly known as ***twin Earth.*** It is the hottest planet in the solar system. It spins in a direction opposite to that of its rotation around the Sun, referred to as ***retrograde*** motion. Venus takes longer to spin once about its axis than to rotate once about the sun. In other words, a day (243 earth days) on Venus is longer than the year (225 earth days). It is the slowest planet in the solar system and hence has a near spherical shape. Venus is the brightest night sky object next only to the Sun and the Moon. Venus can be seen immediately after sunset and is known as the ***evening star.*** It can also be seen as a bright daylight sky object before sunrise and hence is called the ***morning star.*** Venus holds a blanket of atmosphere round it just like the Earth. Atmosphere of Venus consists of carbon dioxide and nitrogen but notably no water vapor. Carbon dioxide, while being transparent to visible light, blocks the infrared rays re-emitted by the surface of Venus. This greenhouse effect raises the surface temperature of Venus to a scorching 460°C, which can melt even lead. There is a cloudy layer of sulfur dioxide in the upper atmosphere of Venus. This results in a sulfuric acid rain, which does not reach the surface of Venus but evaporates at a height of around 25 km in the atmosphere. When perfectly aligned, Venus passes in front of the Sun and appears like a small black dot moving across the disk of the sun. This phenomenon is referred to as a *transit event*. Mercury also gives rise to a transit event.

Earth is the only planet that is not named after a god. Earth is estimated to be 4.543 billion years old. It is also the densest planet of the solar system. The Moon is the only natural satellite of the Earth. Earth takes 365.25 days to complete one rotation around the sun. The extra 0.25 or a quarter day is added up to an additional day in the month of February once every 4 years in the ***Leap Year***. Light from the Sun takes 8 minutes and 20 seconds to reach the Earth.

What is remarkable about the Earth is it is the only habitable planet in the solar system. Habitable zone around a star is defined as the region in which a planetary surface possesses liquid water that is essential for supporting all forms of life on the planet. Almost 75% of Earth's surface is covered with either liquid water or frozen water. Because of this unique feature, Earth has rightfully earned the title ***'Blue planet'***. In addition to this vast water resource, the Earth has several attributes that make it the ideal place for evolution of all life forms, including intelligent life. Earth is placed at an ideal distance from the parent Sun, which has enabled it to retain its oceans in the liquid form. Closer to the Sun, water would evaporate leaving a desert behind, as with Mars and Venus, and any further away from the Sun all the water would freeze. Geoffrey Marcy, an astronomer at the University of California, Berkeley who has played a key role in the discovery of dozens of extrasolar planets, has this to say: "The most impressive attribute of the Earth is the existence and amount of liquid water on its surface". In addition, plate tectonics of the Earth-slipping and sliding of the plates in the outer shell of the Earth-are not only responsible for creating the mountains and oceans on the Earth's surface, but also in moderating the carbon-silicate cycle over geological time scales, and in regulating carbon levels in the atmosphere to sustain the temperature required to maintain water in the liquid form. The Earth also has the right size to hold water on its surface. A much smaller Earth would fail to hold on to a similar atmosphere and a much larger Earth would end up as a gas giant, too hot a place to support life. Our big brother, Jupiter, keeps blocking much of the undesirable debris like rocks as small as cars and as huge as moons sniffing life out of our planet (Hamblin and Christiansen 2003). Our friendly moon also helps stabilize the Earth's rotation, preventing its poles from moving erratically. In all, the Earth is pitched at the precise distance from the Sun, with the right tilt of its spin axis and with the right rotational period for evolution of life. Whether this precision is a mere stroke of luck or is a possibility at other places in the universe with equally favorable conditions is the subject matter of this book.

Mars, also known as the ***Red planet*** because of its appearance, is named after the Roman ***god of war.*** The Reddish color of the planet is due to the presence of iron oxide in its soil. It is half the size of the Earth and has an atmosphere composed mainly (95.3%) of carbon dioxide. Atmosphere of Mars is very dry since all water on its surface has frozen to ice. The most conspicuous feature of Mars are the polar caps composed of a thin layer of frozen water and frozen carbon dioxide or ***dry ice.*** Mars has two irregular shaped moons: ***Phobos*** and ***Deimos.*** Mars is free of clouds but has occasional dust storms, which etch out long cracks on its surface. Proximity to the Sun raises the temperature of the planet, enabling the loose dust particles to rise and spread out.

Millions of asteroids made of rock and metal orbit the sun in the ***asteroid belt*** between Mars and Jupiter. Largest asteroid is ***Ceres*** with a diameter of around 760 km. The asteroids are composed of mostly hydrogen and helium with a relatively small rocky core.

Mars is at the edge of the habitable zone around the Sun. As per an ancient theory, Mars formerly was in the habitable zone but due to some catastrophic event, it is no-longer habitable.

Jupiter, Roman ***king of gods,*** is the largest and fastest planet of the solar system and the most massive at more than 300 times the mass of Earth. Jupiter has 69 moons of which ***Io, Europa, Ganymede*** and ***Callisto***

were discovered by Galileo in 1610. Ganymede is the largest moon in the solar system. There is a prominent oval-shaped spot in the southern hemisphere of Jupiter, which has been visible ever since its discovery in 1664 by Robert Hooke. This spot is supposedly a huge tornado measuring around 14,000 km wide and nearly 40,000 km long with a deep brown tan. German astronomer Ernst Temple aptly named it the *Great Red Spot*. Spectral analysis in the optical and infrared wavelengths reveal the Jovian atmosphere to be composed mainly of hydrogen and helium in very much the same proportion as found in the sun. If Jupiter had eight times more mass, it could have ended up as a star! Being gaseous, Jupiter shows a differential rotation just like the sun. In 1979, the Pioneer probes revealed that Jupiter emits twice the amount of radiation that it receives from the Sun, suggesting an internal source of heat. It is surmised that Jupiter is still in a phase of cooling off (Joshi and Rana 2011).

Saturn, Roman ***god of harvest,*** is the second largest planet of the solar system. Saturn's density is less than that of water. It would float in a swimming pool on Earth! Like Jupiter, the atmosphere of Saturn also is composed mostly of hydrogen and helium. Saturn also gives out 1.8 times the radiation it receives from the Sun. The ring system around Saturn was discovered by Galileo in 1610 and is familiar to every amateur astronomer. Edward Roche, a French astronomer and mathematician, proposed that any satellite in orbit within 2.44 times the radius of a planet (referred to as *Roche limit* or *Roche radius*) should disintegrate into small particles due to tidal forces from the planet and form a ring of debris around it. Saturn has 18 moons of which **Titan** is the largest, slightly bigger than our moon. It is the only satellite in the solar system with no atmosphere around it.

Uranus, Greek ***god of sky,*** is the third largest planet by diameter. It is also a lopsided planet like Venus which rotates from east to west, i.e. anticlockwise as seen from the North Pole of the Sun. In addition, it spins about an axis which is tilted by 98 degrees, much like a bottle rolling on the ground!!!

Neptune, Roman ***god of the sea,*** is the farthest, coldest and the slowest planet with the longest year. One year on Neptune is equal to 165 Earth years!!! Position of Neptune was first predicted based on the theory of gravitation before it was discovered by Johann Galle in Germany in 1846. Since its discovery, Neptune just completed orbiting the sun once in 2011. Neptune is the fourth largest planet by diameter and the densest gas giant. Neptune has characteristics similar to those of Uranus and Jupiter; it has a great dark spot similar to Jupiter almost at the same latitude.

2.4 Earthlings in the Cosmic Order

Our understanding of the universe at large is inadequate for earthlings to lay claim to being the only intelligent race in the cosmos. Several stars like our Sun along with their planetary companions have been discovered. But the question still remains: are there human species in other parts of the universe? If yes, how are we going to find out and once found, what will be our mode of communication with these so called 'aliens'? Nevertheless, discovery of humans in any form anywhere other than the Earth would be a sensational breakthrough.

Frank Drake made the first attempt in 1960 to search for extraterrestrial life through *'Project Ozma'*. Drake sent radio signals to two nearby stars hoping to get a response of some sort but got none. He later took over as the director of the then world's largest radio telescope, a 1000 feet dish at Puerto Rico. He beamed a radio signal again in 1974, which could be received by a similar telescope in any part of the Milky Way galaxy but he was unlucky for the second time. Sadly, on 1st December 2020, this mammoth telescope housed at Arecibo observatory had a disastrous fall from a platform supporting it at a height of 450 ft above the ground, weeks after it was declared that the telescope would be dismantled for safety reasons!

The following words of Lord Tennyson poetically sums up the travails of this amazing universe:

This world was once a fluid haze of light,
Till towards the center set the starry tides,
And eddied into suns, the wheeling cast
The planets, then the monster, then the man.

Albert Einstein, a genius theoretical physicist of the 20th century, once very fittingly remarked, *"The most incomprehensible thing about the universe is that it is comprehensible!!"*

2.5 Possibility of Life Elsewhere in the Solar System

Astrobiology is the study of the origin and evolution of life in the universe. It deals with identifying habitable environments within the solar system and outside of it, and then looks at essential ingredients for supporting life in them. A habitable planet should be basically rocky and have a reservoir of water, which is capable of dissolving basic nutrients for organisms to consume and grow. Elements necessary for the chemistry of life are carbon, oxygen, hydrogen, nitrogen, phosphorus and sulfur. All life forms rely on some source of energy for metabolism to help them survive and sustain. This energy mostly comes from the parent star, similar to the Sun for our Earth.

Since its establishment in 1958, NASA began its journey towards looking for life within the solar system and elsewhere in the universe.

Europa is one of the smallest moons of Jupiter. It became a candidate with a habitable environment (Nordheim et al. 2017) since NASA's Galileo spacecraft collected evidence of a magnetic field created in Europa as Jupiter's powerful magnetic field swept across it. This magnetic signature could be due to an ocean of salty water. 'Ridges' and 'grooves' on the bright icy surface of Europa could transport subsurface water to the surface (Thompson 2018). Life on Europa could exist under an ice ocean, which is similar to our Earth's deep-ocean hydrothermal vents. Scientists believe that essential chemicals were present in Europa even as it formed. Later on, impacting asteroids and comets would have deposited more organic materials onto Europa. However, there is no possibility of any insolation on the icy moon Europa. So life on Europa may not be powered by photosynthesis but through some chemical reactions!

Enceladus is an icy moon of Saturn and is the sixth largest moon in the solar system. It is also the most reflective object in the solar system (Jet Propulsion Laboratory 2005). Cassini spacecraft's close flybys revealed the existence of atmosphere on Enceladus and also recorded its unmistakable magnetic signature. Magnetic field oscillations clearly pointed to the ejection of plumes of salted water from underneath, laced with grains of silica-rich sand (Tobie 2015). In contrast to studies of the potential habitability of Jupiter's moon Europa, the warm oceans of Enceladus may be habitable if they are chemically reacting with the overlying ice shell on time scales spanning over millions of years (Christopher et al. 2008). Since 2015, researchers have been wondering if the warm oceanic water could be interacting with rocks to generate a form of chemical energy suitable for some forms of life to use and evolve. Scientists are also of the opinion that Enceladus is too young for any form of life to show up as of now.

Titan, orbiting Saturn, is the second largest moon of the solar system, larger than even planet Mercury. Cassini spacecraft's gravity measurements revealed an ocean of liquid water possibly mixed with salts and ammonia beneath the icy floor of Titan. In addition, Titan's lakes and seas of liquid methane and ethane may provide a habitable environment for forms of life other than those on Earth in its subsurface oceans and in the hydrocarbon liquid on the surface. For possible life on

Titan, it is necessary to understand Titan's current atmosphere of methane and nitrogen, which is of biogenic origin, and explore it from an angle of potential productivity of oceanic microbes. Even though microbial life appears to be possible, it is unlikely that Cassini or Huygens missions will be able to distinguish these unambiguous biomarkers. Possibility of Venus being habitable is discussed in Section 11.10.

CHAPTER 3

Exoplanets

'That is a big question we all have: Are we alone in the universe? And exoplanets confirm the suspicion that planets are not rare.'

– Neil DeGrasse Tyson

3.1 Introduction

Earth is undoubtedly a special planet in the solar system with its unique features capable of supporting life. But is Earth really that special? Is our world unique? Are we alone in this vast universe? Is life nurtured in planets in orbit around other stars? Exoplanets are planets in orbit around their parent stars outside the solar system. Study of exoplanets and their habitability has been a very fascinating area of research in recent years. With improving technology, planet hunters are confident of discovering the Earth's siblings. These studies can give valuable insight into answering questions like: are there planets, perhaps a large number of them, hidden in the darkness of space harboring living species, even intelligent ones, more intelligent and more advanced than us, but completely unknown to us? The probability of existence of complex life in an exoplanet was proposed by Kasting (1993). Gaidos et al. (2005) recommended the right mass range for exoplanets with a sustainable atmosphere necessary for supporting life to be between 0.1 to 10 times that of the Earth. Corresponding range of radius for a habitable exoplanet is from 0.5 to 2.5 Earth radius and a temperature range from 182 to 285 K.

Discovery

The efforts to look for extraterrestrial intelligence were initiated nearly a century ago. The real breakthrough was in 1995 when the first exoplanet, named 51 Pegasi *b* (Mayor and Queloz 1995) was discovered approximately 50 light years away in the constellation of Pegasus (Figure 3.1) orbiting around the main sequence host star 51 Pegasi.

FIGURE 3.1 Artist's impression of the 51 Pegasi *b* (Credits: ESO)

Presently, focused efforts have been devoted by several space missions to the discovery of habitable worlds. The first spacecraft CoRoT mission (Convection, Rotation et Transits planetaires) CoRoT mission by France, launched in 2006, was followed by the most popular launch of Kepler spacecraft by NASA (2009), and the recent TESS (Transiting Exoplanet Survey Satellite). While CoRoT has discovered 24 exoplanets, NASA's Kepler has confirmed 25 exoplanets and 1235 candidates. Future ventures like James Webb Space Telescope mission scheduled for launch in March 2021 and Planetary Transits and Oscillations of stars (PLATO) to be launched in 2026 will aid humankind to understand the climatic conditions of a large number of exoplanets and to verify certain hypotheses (Carone et al. 2016) with regard to exoplanetary atmospheres. To date, more than 4000 exoplanets have been discovered and are regarded as 'confirmed'. There are thousands of other candidates waiting out there in space to be explored and discovered.

3.2 Nomenclature of Exoplanets

The exoplanet nomenclature is an extension of the system used for naming multiple-star systems as adopted by the International Astronomical Union (IAU). An exoplanet orbiting a single star is named after its parent star followed by a lowercase letter, as in the case of 51 Pegasi *b*. Parent stars are designated with a letter in the upper case. Letter *'a'* is not used to designate exoplanets. Letter *'b'* signifies it was the first exoplanet discovered in orbit around the host star. If other planets are found around Pegasis, their names will be followed by the letters *c, d, e, f,* etc. specifying the sequence of their discovery, but not their orbital placement around the host star. It is, however, not essential that an exoplanet be found orbiting around only a single star. Some exoplanets have two Suns or a binary system of parent stars. The brighter and hence the larger star of a binary system is designated with upper case *A* and the other as *B*, for example Sirius *A* and *B* in the constellation of Canis Major. If an exoplanet orbits around only one star of the binary system as shown in Figure 3.2, it is said to have an '*S*-type' orbit. The orbit is said to be '*P*-type' if the exoplanet goes round both the stars and is called a circumbinary planet.

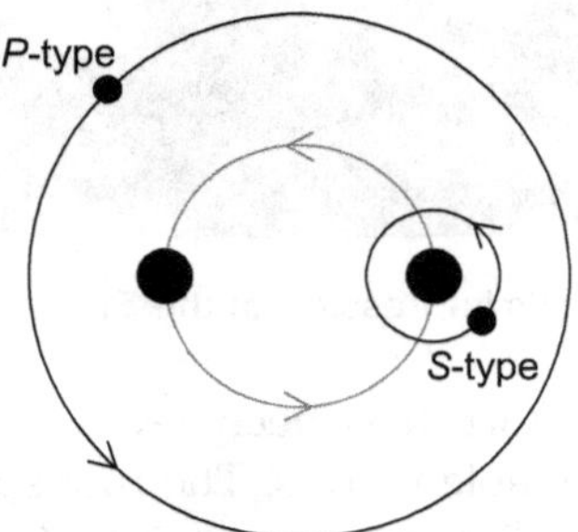

FIGURE 3.2 *S* and *P* orbitals of exoplanets

For example, Kepler-34 (*AB*) *b* is the first exoplanet discovered by NASA's Kepler space telescope in 2012,orbiting around the parent binary star Kepler-34 (*AB*) in the constellation of Cygnus, about 4890 light years away. Number 34 following 'Kepler' indicates the order of detection of this binary star system as per the instrument's data. '*Ab B*' and '*A Bb*' refer to exoplanets in *S*-type orbit around the brighter star and the dimmer star of a binary system, respectively. Alpha Centauri is a triple star system and an exoplanet discovered around the 2nd brightest star in the system is named Alpha Centauri *Bb*.

3.3 Observational Techniques

An exoplanet is a very faint source of light as compared to its host star. This makes it very difficult to directly observe an exoplanet, let alone resolve it from its parent star. For instance, the Sun is around a billion times as bright as the light reflected from any of the objects in the solar system. Astronomers have resorted to various indirect techniques to actually differentiate an exoplanet from its bright host star: Radial Velocity Technique, Transit and Occultation Photometry, Microlensing, Direct Imaging etc. (Seager et al. 2007, Seager and Deming 2010, Anglada-Escud et al. 2013, Sengupta 2016), the details of which are presented in Chapter 5. The parameters such as mass, radius, temperature and escape velocity of exoplanets are arrived at by employing these techniques. A detailed comparative study of these parameters with those of the Earth helps in the understanding of their potential habitability.

3.4 Host Star

Host star or extrasolar star is the name given to the parent star of an exoplanet. For example, 51 Pegasi is the host star for the exoplanet 51 Pegasi *b*. A host star is at the center of a planetary system, which gravitationally holds a host of planets, irrespective of their size and the tiniest particles of debris bound to it. This gravitational force provides the centripetal force necessary to keep the planets orbiting around the host star. To find life on other planets, the host star's stellar type plays a major role and many observations are done around our Sun-like *G*-type stars and also much cooler stars of *K* and *M*-types.

3.5 Exomoons

Exomoons or extrasolar moons are natural satellites, similar to our moon that orbit around an exoplanet or any other extrasolar body which is not a star. With the currently available technology of space observations, it is extremely difficult to detect exomoons because of their enormous distances from the Earth as also their small size. To date, several exomoon candidates have been detected. Some of the exomoon candidates detected by the Kepler space telescope suggest they might be interesting goldilock zones capable of sustaining extra-terrestrial life forms. It is too early to make any conclusive statements since the search for exomoons is still in its infancy.

Kepler-1625b, a Jupiter-sized gas giant, 8000 light years away, in orbit around its host star was discovered in 2016. In 2017, researchers found a Neptune-sized exomoon at a distance of 20 planetary radii orbiting Kepler-1625b.

3.6 Hot Jupiters

About 1% of the exotic bonanza of exoplanets belong to a unique class called Hot Jupiters. Hot Jupiters are giant, gaseous exoplanets that have mass comparable to Jupiter of the solar system, but much hotter and have shorter orbiting periods of less than 10 days around the host star! The short orbital period indicates close proximity of the hot Jupiter to its host star (less than 0.1 AU) in contrast with the gas giants Jupiter, Saturn, Urnaus and Neptune, which are located in the outer orbitals of our solar system (Pollack et al. 1996). This physical closeness to the host star, in turn, implies a high surface temperature for the Hot Jupiter.

In 1995, Michel Mayor and Didier Queloz reported the detection of the first hot Jupiter, 51 Pegasi *b* (Wang and Fischer 2015) in orbit around its host star 51 Pegasi, using radial velocity technique. Of the 4000 exoplanets confirmed till date, around 350 could belong to the class of Hot Jupiters.

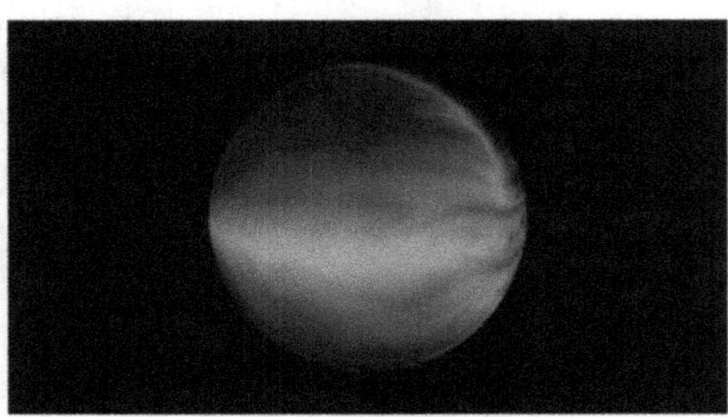

FIGURE 3.3 The turbulent atmosphere of the hot gaseous planet HD 80606b, from NASA's Spitzer Space Telescope (Credits: NASA/JPL-Caltech/MIT/Principia College)

HD 80606b is an eccentric gas giant with a mass of 4 Jupiters in the constellation of Ursa Major at a distance of 190 light years away. It orbits its host star HD80606, which is a member of a binary system, its close knit sibling being HD80607. Uniqueness of HD 80606b lies in its extremely elongated orbit, in fact the highest known till date, comparable to that of comet Hailey in our solar system. Perihelion of HD 80606b is 13 times closer than the distance of Mercury from the Sun. At aphelion, it is more than 25 times this distance! Proximity to the host star causes fierce storms on the planet as is evident in Figure 3.3.

General Characteristics of Hot Jupiters

1. Have large masses (0.36 to 11.8 Jovian mass) and short orbital periods ranging from 1.3 to 111 Earth days (Winn et al. 2010). Researchers have found that predominantly oversized Hot Jupiters orbit around older host stars. For instance, Hot Jupiters HAT-P-65b and HAT-P-66b discovered with HATNet (Hungarian-made Automated Telescope Network) in Arizona and Hawaii are around 5.46 billion and 4.66 billion years old, respectively, and their host stars have completed 84% and 83% of their life spans.
2. Most have nearly circular orbits. The reason for circular orbits could be perturbations from the nearby stars or tidal deformation (Fabrycky and Tremaine 2007).

3. Hot Jupiters are tidally locked to their host stars (Močnik et al. 2016). Tidal forces come into effect when the gravitational pull on a large object is more on one side than the other. This will cause the planetary object to stretch and get tidally bound to the host star. The exoplanet now enters into a synchronous rotation around the host star, with its same side facing the star.
4. Many Hot Jupiters have unusually low densities. For example, HAT-P-65b and HAT-P-66b have masses of 0.5 to 0.8 times the mass of Jupiter, but have diameters 1.9 and 1.6 times that of Jupiter, respectively. This suggests that these Hot Jupiters are unusually puffed up, suggestive of a very low density. The lowest measured density thus far is that of the TrES-4 exoplanet at 0.222 g/cm^3 (Mandushev et al. 2007).
5. Hot Jupiters appear to be more common around *F*-type and *G*-type stars with their surface temperature of 5000 to 7500 K, respectively, and less so around cooler *K*-type stars with surface temperature ranging from 3500 to 5000 K.

It is mysterious how the Hot Jupiters get so close to their host stars. Several theories have been put forth to account for this. Three main theories of hot Jupiter formation are:

1. *In Situ formation*: According to this theory, a Hot Jupiter can form in the proximity of its host star if a fraction of the protoplanetary disk fragments can bind together or a rocky, primitive super Earth (with several times the mass of Earth) can accrete gaseous matter several times its own mass from the protoplanetary disk (Dawson 2018). But it can be shown that neither of the above two processes is practically plausible.
2. *Gas disk migration*: It is suggested that the torques arising in the gaseous protoplanetary disk can shrink a giant planet's semi-major axis to a mere hundredth of its value. This may drive the giant planet to migrate inwards at a rate dependent on opacity, viscosity and entropy profile of the disk. Migrations that are fast in comparison with the lifetime of the disk can compel the host star to engulf the planet. This theory of formation of Hot Jupiters is still under debate.
3. *Tidal migration*: It is believed that giant planets with highly eccentric orbits can undergo tidal migration towards their host stars. This involves two stages of reducing the planet's angular momentum and energy. In the first stage, a perturber in the form of another planet robs the giant planet's angular momentum and renders its orbit more elliptical. During the second stage, the giant planet dissipates its energy tidally with the host star and approaches it, till it settles

down in a near circular orbit. Both these stages seem plausible from the observed eccentricity of some Hot Jupiters.

3.7 Goldilocks Zone

It is not possible for life to evolve on all exoplanets. A planet too close to the host star will be far too hot, while a planet farther away will be too cold and icy to sustain life on it (refer Figure 3.4). In astrobiology, the term 'Goldilocks Zone', also variously called as circumstellar habitable zone (CHZ), 'liquid water belt', 'ecosphere' or 'life zone', is used to specify a region around a star where a planet with sufficient atmospheric pressure can retain water in liquid form on its surface. In other words, the temperature of a planet in the life zone typically lies between the freezing point (0°C) and boiling point (100°C) of water. While Earth is in the habitable zone of the Sun, Mars is at the very edge of this zone.

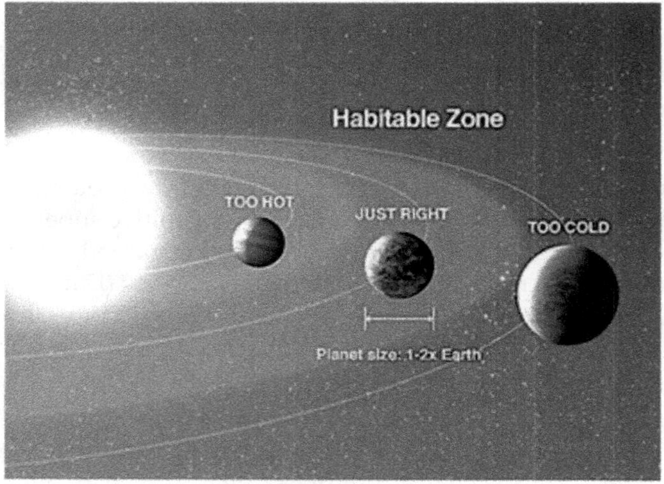

FIGURE 3.4 Habitable zone around a star (Credits: NASA)

Surface planetary habitability preempts the planet to be orbiting around the host star at the precise distance, in addition to several geophysical and geodynamical considerations, the environment of the host star, its luminosity and age, and long-term photosynthetic biomass production (Cuntz and Guinan 2016). A potentially habitable planet with Earth-like conditions will favor Earth-like life forms.

3.8 Rogue Planets

Rogue planet, also known as a wandering planet, free-floating planet, nomad planet or an interstellar planet is a planet without a host star that drifts through intergalactic space or directly orbits around the galactic center. Refer Figure 3.5 for an artist's impression of a rogue planet.

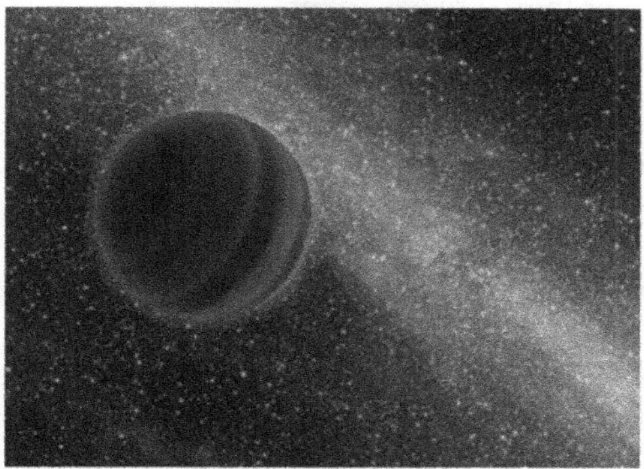

FIGURE 3.5 An artist's impression of a rogue exoplanet (Credit: NASA/JPL-Caltech)

These homeless planets, so to say, have no sunrises and sunsets. How common are these rogue planets? As per a study conducted in 2011, in our own Milky Way, it is estimated that 50% more rogue planets exist than planets with host stars. In other words, the Milky Way galaxy alone could be hosting millions of these rogue planets. But how do these planets form in the first place? There are many theories to explain why a planet goes rogue. As per one theory, untold things can happen in the neighborhood of two stars or other gravitationally bound massive objects either when they pass each other or interact with each other. Planets in a system can get pulled into lower orbits as with migratory Hot Jupiters or get flung out into space from the planetary system with escape velocities as rogue planets. Planets are also orphaned when their parent star disappears. The sheer force of explosion of a massive star during the end stage of its life as a supernova has the ability to expel planets originally belonging to it into the void of space with high speeds. It is however believed that most rogue planets were formed in the early stages of formation of their host star system. Researchers are more interested in finding out if these rogue planets are habitable. Far out in the frigid depth of space and devoid of a host star, the rogue planets must be extremely cold. But the rogue planet possibly holds on to the residual heat of its formation and perhaps a liquid ocean beneath its icy crust capable of supporting life.

In 2011, Astrophysicist Takahiro Sumi of Osaka University in Japan observed an exoplanet using a microlensing technique. Also, observations of millions of stars in our galaxy using Mount John Observatory in New Zealand, in collaboration with Las Campanas Observatory, Chile, revealed there are at least 2 Jupiter-mass rogue exoplanets around every host star in our Milky Way alone. Life on these rogue planets, however, is highly unlikely as most of them are gas giants like our solar system Jupiter.

Interesting Facts

1. ***Extragalactic planets:*** They are star bound exoplanets or rogue planets outside the Milky Way galaxy. Their very large distances from the host star makes it very hard to detect them. The first extragalactic planet to be discovered in 1999 was around a star in the disc of M31 in the Andromeda galaxy. This detection became possible during the microlensing event of a red giant in the background (Dai and Guerras 2018). The most distant extragalactic planets SWEEPS-04 and SWEEPS-11 at 27,710 light years from the Earth, located in the Sagittarius constellation, were discovered in 2019. There are a myriad of extragalactic exoplanets that are yet to be detected (Dai and Guerras 2018).
2. ***Exocomets:*** Comets in general contain materials in their pure pristine physical and chemical memory of the early stages of formation of the parent star (Strom et al. 2020). A comet sublimates on approaching the host star emitting a jet of dust and gas, known as a comet tail and it points away from the star. Exocomets are comets observed around exoplanets and include rogue comets also. The first ever exocomet was detected around Beta Pictoris (β pic) in 1987 (Ferlet et al. 1987). Due to their volatile nature, comets are also known as Falling Evaporating Bodies (FEBs) (Beust et al. 1990). Exocomets are detected through photometric transits or via variable absorption features in the CaII H or K lines in the far IR emission spectrum in the debris disk. Though spatial resolution of exocomets is not satisfactory, efforts are on to study these commonplace objects in the vicinity of an exoplanet for a better understanding of the formation of exoplanetary systems.

CHAPTER 4

Missions and Observation Facilities of Exoplanets

'To consider the Earth as the only populated world in infinite space is as absurd as to assert that in an entire field sown with millet, only one grain will grow.'

– Metrodorus of Chios

To date, more than 4000 exoplanets have been discovered using the space based facilities and they are backed up by ground based observations. *F, G, K* and *M*-type stars are the targets for discovering habitable exoplanets. Only Kepler-1632b is found to orbit an *F*-type star. *G*-type stars are expected to have habitable planets around them very similar to the Earth. *K*-type stars would host super habitable exoplanets, exomoons rather than exoplanets. *M*-type dwarf stars are the most common ones in the universe since these stars appear dim in the optic range the planets around them are clearly visible but their habitable zones are yet to be clearly studied and understood. 55 potentially habitable exoplanets have been identified as of March, 2020 of which 20 are believed to be terran or Earth-like, while 1 is sub-terran or Mars-like and 34 are super-terran or super Earths.

4.1 CoRoT Mission

Convection, Rotation and planetary Transits (CoRoT) is a mission led by the French Space Agency, CNES (National center for space studies in French), which started to detect exoplanets a few to several times larger than our Earth in 2006. By 2008, CoRoT reported the discovery of two exoplanets of the 'Hot Jupiter' type and improved its exoplanet count to 32 by 2013. Later, the European Space Agency (ESA) joined the mission by providing the optics for the telescope and for testing the payload. After the launch, CoRoT was placed in an orbit that allowed for continuous observations for more than 150 days each of two large regions of the sky located in opposite directions. The payload of the CoRoT satellite consisted of a telescope, two cameras – one each for the two mission objectives (exoplanet search and asteroseismology- study of internal structure of a star by observing the frequency spectrum of its different oscillatory modes), and on-board computer processors (see Figure 4.1). Though we can observe Hot Jupiter sized exoplanets from ground, space is the preferred place for viewing Earth-sized exoplanets. CoRoT discovered many rocky Earth-sized exoplanets and this success story inspired several space missions to search for more exoplanets. The CoRoT satellite, however, stopped transmitting data in November 2012 and had to be withdrawn in June 2013.

FIGURE 4.1 An artist's impression of CoRoT mission (Credit: ESA)

4.2 Kepler Space Mission

In March 2009, NASA mission successfully launched the Kepler Space Telescope (named after the astronomer Johannes Kepler renowned for the laws of planetary motion) to scan the sky in the constellation of Cygnus to discover unknown worlds using the Transit technique (see Figure 4.2). The Field of View (FOV) was adjusted so as to avoid the glaring Sun. By scanning a hundred thousand stars simultaneously in the constellations of Cygnus and Lyra, Kepler was able to detect a number of Earth-sized exoplanets around Sun-like stars in the optical range of 430-890 nm.

The scientific objectives of the Kepler Mission were "to determine the percentage of terrestrial and larger planets that are in or near the habitable zone of a wide variety of stars, distribution of sizes and shapes of the orbits of these planets, estimate how many planets are there in multiple-star systems, determine the variety of orbit sizes and planet reflectivities, size, mass and density of short-period giant planets, identify additional members of each discovered planetary system using other techniques, and determine the properties of those stars that harbor planetary systems".

FIGURE 4.2 An artist's impression of Kepler telescope (Credits: NASA)

The Kepler space telescope discovered thousands of new worlds before it ran out of fuel. By February 2011, Kepler had detected 1235 exoplanets out of which 54 could be in the habitable zone. This tally increased to 2326 by December 2011 of which 207 were Earth-size, 680 super-Earth-size, 1181 Neptune-size, 203 Jupiter-size and 55 larger than Jupiter, and the count reached 3278 by June 2013. The planetary data from the various missions are maintained by the Planetary Habitability Laboratory-Exoplanet Catalog (PHL-EC) @ UPR Arecibo. As per data of research from Kepler's Space Telescope, almost 50% of the stars with temperatures similar to that of the Sun could have rocky planets around them capable of harboring liquid water required to support life.

4.3 TESS Space Mission

The Transiting Exoplanet Survey Satellite (TESS) is NASA's Astrophysics Explorers program designed to search for exoplanets using the Transit method (see: Chapter 5) in an area 400 times larger than that covered by the Kepler mission. TESS is operated by MIT in Cambridge, Massachusetts and run by NASA's Goddard Space Flight Center in Greenbelt, Maryland since 2008. TESS will spend a year surveying the southern sky and then study the northern sky over the next year. In all, TESS telescope (Figure 4.3) will be able to peer into the depths of 85% of the sky, including 200,000 brightest stars that are also close to the Earth.

FIGURE 4.3 An artist's impression of TESS telescope (Credits: NASA)

Interestingly, TESS was launched by Elon Musk's SpaceX rocket company, which is a private agency pioneering in the current space race.

4.4 Ground Based Observatories

More than 35 ground based observatories of extrasolar planets are currently active. The famous ground observatory of exoplanets is WASP or Wide Angle Search for Planets, which has discovered many exoplanets-the highest. WASP is an international consortium conducting an ultra-wide angle search for exoplanets using transit photometry since it became operational in collaboration with Queen's University and University of St. Andrews. The array of robotic telescopes aims to survey the entire sky for planets, simultaneously monitoring many thousands of stars with apparent magnitude ranging from 7 to 13. Most of the exoplanets are named after the telescopes which detect them or confirm their presence.

CHAPTER 5

Detection Techniques and Data Archives of Exoplanets

"Shouldn't we be content to be cosmic sloths, enjoying the universe from the comfort of Earth? The answer is, no."

– Stephen Hawking

Exoplanet is a life harboring planet external to our solar system. The search for exoplanets and life on them was initiated a few decades ago. But the systematic scientific approach towards this began in 1995 with the discovery of the first confirmed exoplanet orbiting around 51 Pegasi. Since then, many exoplanets have been discovered using one or more techniques simultaneously (Mayor and Queloz 1995). Some of the methods popularly used by astronomers to detect exoplanets are Radial Velocity or Doppler Method, Transit and Occultation, Gravitational Microlensing and Direct Imaging (Seager et al. 2007, Jones 2008, Seager and Deming 2010, Sengupta 2016).

5.1 Doppler Spectroscopy

Doppler shift technique, also known as Radial velocity method, is one of the most popularly used methods to detect exoplanets and to estimate their physical parameters (Sengupta 2016). Doppler spectroscopy is the most effective technique for locating exoplanets. It works on the principle of Doppler shift (see Box 1) due to a moving source of light. An exoplanet and its host star gravitationally keep tugging at each other and hence both orbit around their common center of mass called barycenter, which is situated along the line joining their centers of mass. If the host star is very massive in comparison with the exoplanet, the barycenter is located within the star or very close to it. As a result, the star wobbles and its center of mass moves along a tiny circle or an ellipse about the barycenter.

Analysis of light from a star reveals a line absorption spectrum in the foreground of its continuous emission spectrum, similar to Fraunhofer lines in the solar spectrum. Each line is the chemical signature of an element present in the atmosphere of the star. Wavelength corresponding to each of the elements can be measured in the laboratory. Star's wobble gives rise to Doppler shift of the absorption lines in its spectrum. Mere detection of the Doppler shift in itself confirms the presence of an exoplanet.

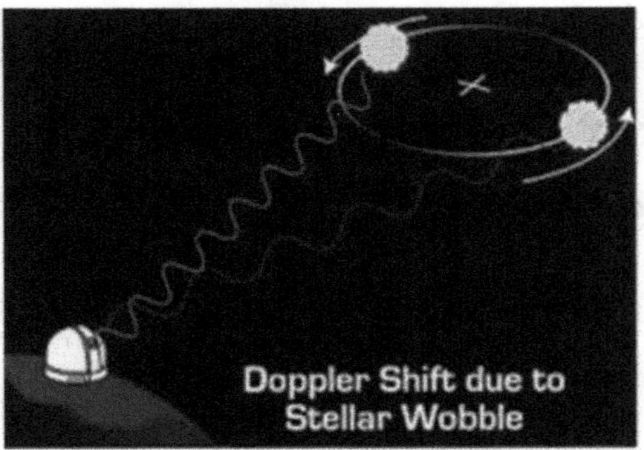

FIGURE 5.1 Red and Blue Doppler shifts due to relative motion of a light source (Credit: Sengupta 2016)

Doppler shift, also known as Radial velocity technique (Figure 5.1), is ideal for locating low-mass, terrestrial (rocky) planets orbiting around their host *M*-type stars. More than 450 exoplanets have been detected employing this method (Schneider 2011), out of which 30 are confirmed while the rest are candidates.

Box 1. *The Doppler effect in sound is familiar to all of us. When a car sounding its horn approaches us, we feel as if the sound is getting shriller (shrillness is directly related to pitch or frequency of the sound waves received by our ears) and the pitch keeps decreasing as it moves away from us. In other words, the wavelength of sound decreases as the source of sound moves relatively towards the observer and increases as the source of sound moves away. A similar effect can be seen with light received from a source as its relative distance from an observer changes. A spectral line with an apparent increase in wavelength is said to be Red-shifted while a line with decrease in wavelength is said to be Blue-shifted. Red shift indicates that the source of light is moving away from us while Blue shift indicates the source is approaching us.*

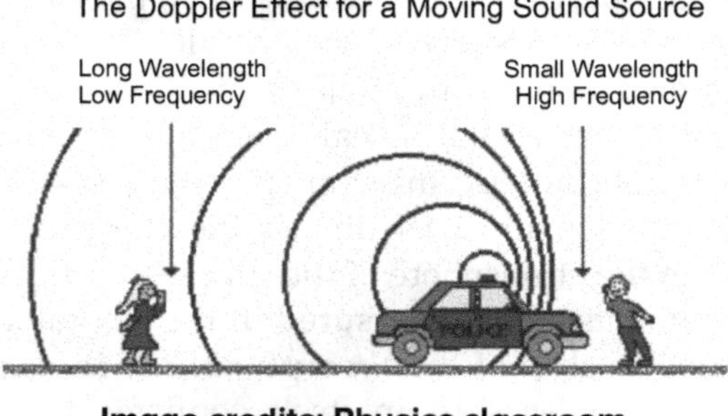

Image credits: Physics classroom

Knowing the lab measured wavelength or rest wavelength of a given line and its Doppler shift, radial velocity of the star (i.e. rate of change of its distance along the line of sight) can be calculated using the formula

$$(\lambda - \lambda_0)/\lambda = \Delta\lambda/\lambda = v/c \quad (5.1)$$

where $\Delta\lambda$ is the Doppler shift in the absorption line of rest wavelength λ_0, λ being its Doppler shifted wavelength, v is the radial velocity of the host star and c is the speed of light in vacuum. A positive value of v indicates the star is moving away from the observer. It also implies $\Delta\lambda$ is positive or the wavelength has increased, and the spectral line is said to be red-shifted. For a star approaching the Earth, the corresponding spectral line is blue-shifted.

Worked Example

The most intense emission line of hydrogen gas occurs at a wavelength of 656.3 nm. The same line is observed to have a wavelength of 660.0 nm in a distant galaxy. Is the galaxy approaching or receding from the Earth? How fast is the galaxy moving?

Solution: Using the Doppler equation $= \dfrac{\lambda - \lambda_0}{\lambda} = \dfrac{v}{c}$

$$\frac{660.0 \text{ nm} - 656.3 \text{ nm}}{656.3 \text{ nm}} = \frac{v}{3 \times 10^8 \text{ ms}^{-1}}$$

$$v = 1.691 \times 10^6 \text{ ms}^{-1} \text{ or } 1691 \text{ kms}^{-1}$$

Since v is positive, the galaxy is moving away from the Earth with a speed of 1691 km per second.

The radial velocity of a star is related to the projected mass of the planet through the equation:

$$v = (2\pi G/P)^{1/3} [M_p \sin(i)/M_S^{2/3}] \text{ for } M_P \ll M_S$$

where M_P and M_S are the mass of the planet and the host star, respectively, P is the orbital period of the planet, i is the orbital inclination angle-angle between the orbital plane of the exoplanet and its equatorial plane, and G is the gravitational constant. Knowing the radial velocity of the host star, mass of the exoplanet can be calculated using the above formula.

5.2 Transit Photometry Technique

> **Box 2.** *Transit in astronomy refers to the phenomenon when a celestial body moves between a larger celestial body and an observer, usually on Earth. The transiting body looks like a tiny dot moving across the disc of the larger body.*

When a planet transits (see Box 2) between a star and an observer, the star's apparent brightness decreases. If this dimming of the star is periodic, it is suggestive of a planetary object orbiting it.

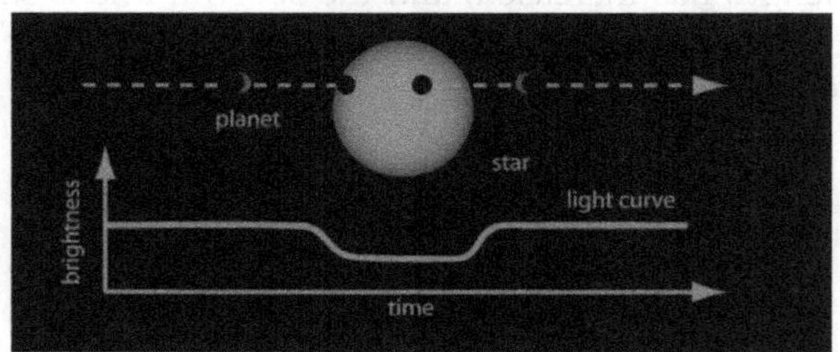

FIGURE 5.2 Transit of exoplanet (Image credit: NASA)

Variation in the apparent brightness of a star implies the presence of a planet (Figure 5.2) in orbit around it. The ratio of decrease in luminosity (ΔL) to its actual luminosity (L) or fractional decrease in the luminosity of the star is given by the relation $\Delta L/L = R_p^2/R_S^2$ where R_P is the radius of the planet and R_S is the radius of the star. The radius of a planet can be determined accurately using this technique if the size of the star is known. Knowing the mass of the planet from the Radial Velocity method and radius from Transit technique, density of the planet can be calculated. Density, in turn, suggests if the planet is rocky or gassy or of some intermediate composition. However, the transit technique is suitable when the orbit of the planet is inclined at a large angle so that it can be viewed edge-on.

Variation in the brightness of a star during transit is between 0.01% to 1% depending on the relative sizes of the exoplanet and the host star. Smaller planets give rise to smaller dips, while the larger planets contribute to a larger dip in the light curve of the host star. Kepler and CoRoT space missions have sensitive instruments to detect this minute dip in the light curve of a star with respect to time. The Kepler mission has discovered over 1000 exoplanets using this method.

Transit technique continues to be the most popular one for exoplanet exploration for several reasons. (i) It is suitable for space-based as well as for ground-based observations of over 100,000 stars at a time. Space telescopes can be directed to view a particular portion of the sky for several months at a stretch. Ground-based observations can be done with even small telescopes as by the 2 telescopes in TRAPPIST (Transiting Planets and Planetesimals Small Telescope), 7 telescopes of HATNet (Hungarian Automated Telescope Network), 2 telescopes of MEarth (pronounced Mirth) project and the 5 telescopes of SPECULOOS (Search for habitable Planets Eclipsing ULtracOOl Stars) survey. (ii) If the planet has an atmosphere, it shows up as dark absorption lines in the spectrum of star light. Through careful spectral analysis during and after transit, it is possible to diagnose the elements in the planet's atmosphere. (iii) During secondary transit, the planet is completely hidden behind the star. By studying the stellar spectrum during and after transit, it is possible to arrive at the planet's color and hence its surface temperature. (iv) The Transit method is ideally suited for observing exoplanets that orbit close to their host stars enabling several periodic transits to be observed and analyzed.

5.3 Gravitational Microlensing

Box 3. *As per Albert Einstein's General Theory of Relativity, humongous concentration of mass warps the surrounding space-time because of its strong gravity. A massive object like a cluster of galaxies, dense core of a galaxy or even an individual massive star behaves like a gravitational lens. It simply means that light from a distant star or a galaxy passing close to the massive object gets deflected. The gravity of the massive object acts like a lens and this results in a sudden increase in the brightness of a distant star or galaxy as the nearby massive object moves. This intriguing gravitational effect can be observed along the line of sight of an observer on Earth.*

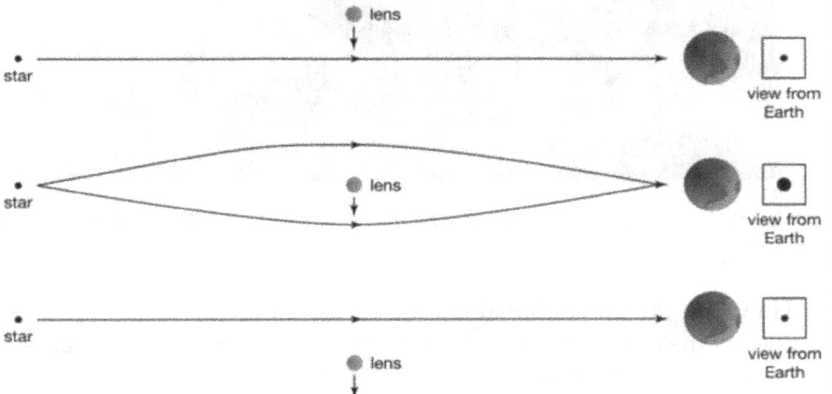

FIGURE 5.3 Gravitational microlensing (Credit: Encyclopedia Britannica)

Gravitational lensing occurs when a massive object behaves like a lens gravitationally influencing the light by a distant star (see Box 3 and Figure 5.3). This visual effect of Gravitational Microlensing results in multiple images of the distant star in the form of thin arcs. This helps astronomers see faint, distant stars and galaxies, and arrive at the distribution of mass in the massive object causing gravitational microlensing.

Gravitational microlensing provides an excellent method of detecting exoplanets. Although the exoplanet has a much smaller mass in comparison with its host star, its gravitational effect on the light curve of the star is fortuitously large and directly observable since large magnifications are caused by the proximity of the planet to its star. It is like decreasing the object distance u, which is inversely proportional to magnification m. Since the process occurs for a period of 2-4 days, the observation cannot be made during the daytime. The event is observed using many telescopes placed at vantage positions in different parts of the Earth.

MOA-2011-BLG-262 refer Fig. 5.4 is a free-floating interesting exoplanet – exomoon system. This was the first microlensing candidate. The exoplanet MOA-2011-BLG-262L could be either a red dwarf or rogue planet and the exomoon MOA-2011-BLG-262 *b* is a Neptune-like planet, with a mass of 17 Earths that orbit the host star at a distance of 0.95 AU with a period of 2.8 years.

FIGURE 5.4 An artist's impression of the MOA-2011-BLG-262 system (Credit: NASA/JPL-Caltech)

However, the gravitational lensing method provides only the ratio of the mass of the planet and its parent star but not the individual parameters as possible with other methods (Sengupta 2016).

5.4 Direct Imaging Technique

Direct imaging technique is based on direct observation of light reflected off an exoplanet or thermal radiation emitted by it. In other words, the method involves taking a direct photograph of the planet. But this method poses difficulty due to the intense glare of the host star particularly when the planet is very close to it. Direct imaging methods are more suitable for planets that are far away from their parent stars and whose orbits can be seen face-on instead of edge-on. This technique is also used to detect rogue planets that are not gravitationally bound to their host stars.

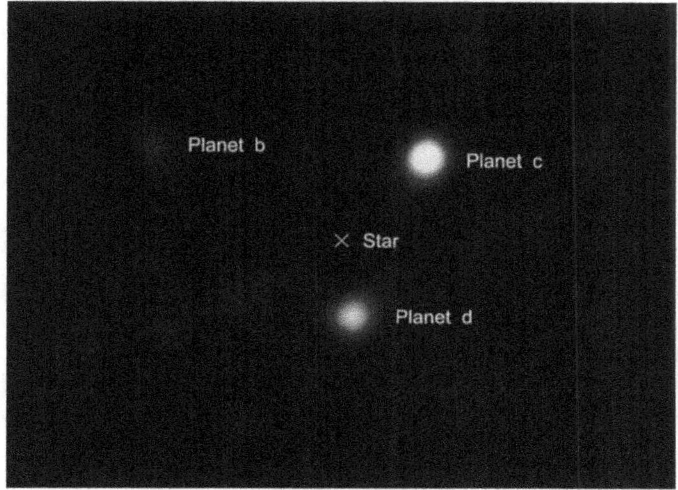

FIGURE 5.5 Direct imaging impressions (Credit: Universe today)

Generally, imaging is done in the infrared since a hot planet radiates more intensely in the infrared than in the visible range of the electromagnetic spectrum. From Figure 5.5, it is evident that detecting planet *b* is difficult since it is very faint as compared to planets *c* and *d*. Direct imaging is used to accurately arrive at the size of the planetary orbit and composition of its atmosphere, specifically the presence of water, methane and carbon dioxide. This method, however, can only give rough estimates of the planet's mass and radius.

Recently, a few exoplanets, which are far from their host stars (distance of 10-30 AU or more), have been imaged directly, by blocking the star light using a Coronagraph (see box 4). But only about 100 exoplanets, around 0.3% of all confirmed exoplanets, have been discovered using this rather difficult technique.

Box 4. *Coronagraph is a telescopic attachment used to view sky objects by blocking the intense glare of a companion star. In the case of the Sun, a coronagraph is used to create an artificial eclipse of the Sun so that its faint corona, nearby stars and planets are revealed in detail.*

Image credit: NASA

5.5 Astrometric Detection Method

Astrometry is the method of detecting exoplanets by measuring the precise positions of their host stars in the sky. The technique basically measures the star's wobble around the center of mass of the planetary system (refer Figure 5.6).

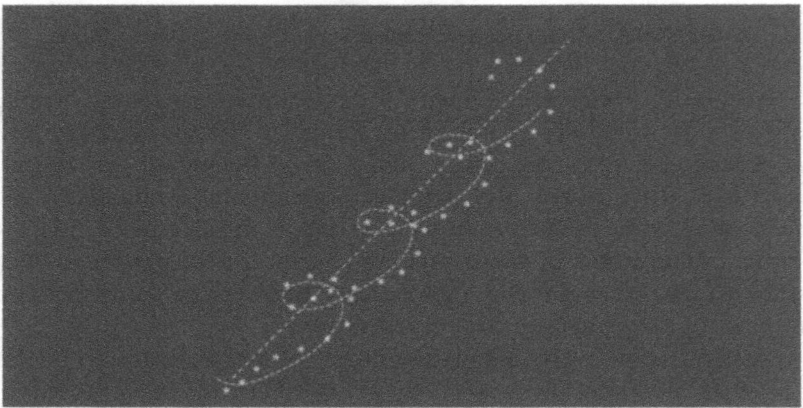

FIGURE 5.6 The observed path of a star across the sky for a period of four years (Credit: ESA)

The motion of the Earth around the Sun makes the star positions to oscillate and each oscillation corresponds to one year. This technique is practically based on follow up observations from missions such as GAIA (Global Astrometric Interferometer for Astrophysics) and ESA (European Space Agency). These missions are expected to discover thousands of planets within a distance of around 500 parsec using astrometric technique. The statistical studies of detected exoplanets are cataloged with respect to different physical parameters to re-confirm their characteristics. This statistical technique helps us to understand multiple-planetary systems.

Data archives: The huge inflow of data from all the techniques used for detection of exoplanets are maintained by many databases. The most popularly used databases are Planetary Habitable Laboratory - Exoplanet Catalog (PHL-EC), NASA data archive, Exoplanet.org and Exoplanet.EU.

CHAPTER 6

Salient Features of Exoplanets

"If a project opens the path for other projects, it means that it has already triumphed!"

– Mehmet Murat ildan

6.1 Color and Brightness

We all know that our Earth looks like a blue globe with swirls of white, a few brown and white patches as seen from outer space. Of all the visible colors of sunlight incident on the Earth, blue color has the shortest wavelength and hence gets scattered more predominantly by the atmospheric particles (Rayleigh's scattering law).

Albedo is a measure of the amount of radiation reflected by a body as compared to the total radiation incident on it. It is possible to determine the color of an exoplanet by its albedo measurements. This was first achieved by the Hubble Space Telescope (HST) for the exoplanet HD 189733b in the constellation of Vulpecula (Evans et al. 2013) orbiting around HD 189733 A. HD 189733b, which is just 63 light years away from us, is an enormous gas giant seething at around 100°C. HST used an Imaging spectrograph to measure the light reflected off the exoplanet-host star system as the exoplanet moved in front and at the back of the host star during its orbit. Interestingly, it was found that the brightness dipped in the blue part of the spectrum whenever the exoplanet was in orbit behind the host star. From this, it was concluded that the color of HD 189733b is blue. Color-color diagrams of planets are used to compare the similarity of exoplanets with the solar system planets. Deep dark blue color of HD 189733b exoplanet (see Figure 6.1) is because of scattering of blue light by silica droplets present in its atmosphere (Evans et al. 2013).

Similarly, helium based exoplanets, which are in majority in our galaxy, are either white or grey in color as revealed by a new theory based on data from NASA's Spitzer Space Telescope.

6.2 Magnetic Field

Magnetic field is an intrinsic property of an exoplanet, which is indicative of the composition of the planet's core, as well as of the dynamical processes taking place in its atmosphere. A rocky planet core indicates rich metallicity, which is the actual source of free electrons inducing the magnetic field. In the case of Jupiter, the interior is made of a rocky core with a gas envelope over it. A combination of a deep-seated dipolar dynamo and a magnetic banding associated with the equatorial jet is a key feature in the prediction of the magnetic field morphology of Jupiter (Gastine et al. 2014). Moreover, as is the case with Earth, magnetic fields, via their shielding effects, play a crucial role in protecting the existent life forms from space radiation and hence support their evolution. Detecting magnetic fields of exoplanets is a difficult task since the only method possible is through observing radio waves arising from the interaction between the host star and the planet's magnetic field.

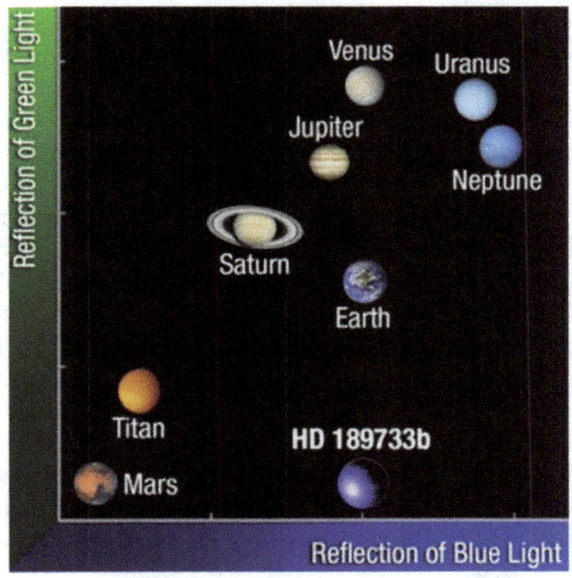

FIGURE 6.1 HD 189733b exoplanet in comparison with solar system planets (Credit: NASA)

HD 189733b is a nearby (63 light years) Hot Jupiter with a ~2 day orbital period. If there was any magnetic activity associated with HD 189733b not disrupted by interaction with the host star's own magnetic field, it should have been detected. There is also a presentation on the state of exoplanet searches in the radio wave range using Low-Frequency Array (LOFAR), Ukrainian T-shaped Radio telescope UTR2 and Giant Metrewave Radio Telescope (GMRT) by Zarka (2010).

Since Hot Jupiters are likely to undergo gradual loss of outer atmosphere with time, it implies that regions undergoing evaporation are prone to have their magnetic fields disrupted. Therefore, the best candidates for radio detection might actually be "cold" Jupiters- Jupiter-like exoplanets at sufficiently large distances from their host stars such that their magnetic fields are not disrupted by the host stars' own fields. It is even better if the cold Jupiter is around stars with strong winds so that its transit is more likely to show up as an independent signal. Ideally, if the exoplanet is a cold Jupiter around a hot star of *O*, *B* or *A*-type, then the magnetic activity from the host itself is rather subdued and can be neglected.

The plot between T_{eff} of the host star against exoplanet temperature shows that there are only 3 cold exoplanets with T~ 0 K and 2 with T~ 500 K which orbit *F*-type stars, which, unfortunately, are known to be magnetically active. If we consider the possibility of a magnetically active star having a strong, magnetically driven wind, then the interaction of the wind with the exoplanet's magnetosphere should give rise to radio emission. However, this emission would then have to be distinguished from that of the host star itself and this might pose some difficulty in detecting it.

On the other hand, it may be easier to look for magnetic activity from exo-planets around stars with strong winds than low magnetic activity, i.e. hot stars. There are 4 planets- Beta-Cir-b, HR3549b, Kepler-13Ab and HD 284149b- orbiting around very hot host stars (A0V to F8V), but with measured effective temperatures of 2000-2500 K. These hardly qualify as cool Jupiters, and the winds of their host stars are not driven by radiation pressure either. Nevertheless, if one really wants to set a search for magnetically active exoplanets, these, or the 3 cold exoplanetary objects mentioned earlier, would be the right candidates. Magnetic field of HD 209458 was first detected indirectly and analyzed by hydrogen gas evaporation (Kislyakova et al. 2014).

6.3 Plate Tectonics

Plate tectonics is the theory that the lithosphere of the Earth, consisting of the oceanic and continental crusts, mohorovicic discontinuity and the upper mantle, is divided into a number of "plates", which move relative to each other across the Earth, similar to huge ice slabs afloat in a sea rubbing against each other. In modern terms, plate tectonics has become synonymous with continental drift-slow and gradual movement of the continents across the Earth's surface over large geological time scales running to millions of years.

Latest research on exoplanets suggests that rocky ones will have Earth-like internal structure. To arrive at this conclusion, Li Zeng and his colleagues at the Harvard-Smithsonian Center for Astrophysics

(CfA) applied the computer-based Preliminary Reference Earth Model (PREM), the standard model for Earth's interior, to accommodate six well established rocky planets with measured masses, size and composition.

According to Van Heck and Tackley (2011), there are two arguments: "The discovery of extrasolar super-Earths has prompted interest in their possible mantle dynamics and evolution, and in whether their lithospheres are most likely to be undergoing active plate tectonics like on Earth, or be stagnant lids like on Mars and Venus. The origin of plate tectonics is not understood properly for Earth yet, and the likely explanation involves a complex combination of rheological, compositional, melting and thermal effects, which makes it challenging to make reliable analysis for exoplanets.

Plate tectonics help in regulating the level of carbon dioxide in the atmosphere of a planet, as it has done with the Earth over ages. In the absence of plate tectonics, carbon dioxide levels build up in the atmosphere leading to a "runaway greenhouse effect", which will effectively trap heat and raise the surface temperature of the planet to uninhabitable levels. For example, Venus, the Earth's twin, has a surface temperature of 470°C, hot enough to melt lead! Although Venus has localized tectonic activity resulting in volcanoes, mountains and valleys, it lacks global tectonics as on Earth, to clear its atmosphere of carbon dioxide.

Plate tectonics have a direct bearing on a planet's climate. Continental drift profoundly affects the distribution of landmass, mountains, oceans, and their connectivity, which in turn controls global climate. Mountains significantly alter rainfall, both locally and globally. They also serve as natural barriers against invasion and migration. Ocean currents regulate global climate and counter the effects of uneven distribution of incoming solar radiation (Insolation).

Based on the work of Cowan and Abbot (2014), if a super Earth has 80 times more water than our Earth, then the entire land will be submerged in the ocean (also known as ocean planets). For a continent to exist, the water content should be less than the aforesaid limit of 80 times less than the Earth.

6.4 Volcanism

Movement of tectonic plates is responsible for the formation of mountains and volcanoes. Gas and dust particles spewed into the atmosphere during volcanic eruptions affect the climate locally. In addition, volcanic material greatly contributes to the fertility of soil and growth of crops for sustenance of different life forms on a planet.

Considerable progress has been made in recent observations of atmospheric signatures of gas giants, but the processes in rocky exoplanets remain largely unknown due to major challenges in observing small planets. According to Demory et al. (2015), large variations in the surface temperature of exoplanet 55 Cancri *e* are observed. This could be due to large clouds of volcanic ash and dust resulting from volcanic activity blocking thermal emissions from the planet.

There is a possibility of large-scale surface activity due to strong tidal interactions between the planet and the host star, or the presence of circumstellar or circumplanetary dust obscuring the planet. This provides a scope for future long-term monitoring of the planet.

FIGURE 6.2 An artist's impression of 55 Cancri *e* in the foreground of its parent star (Credit: NASA/ESA/Hubble Space Telescope)

55 Cancri *e* is the first super-Earth (an exoplanet with mass more than Earth-mass but less than that of ice giants of the solar system) of 8 Earth-mass to be discovered. It orbits with a period of 0.7 days around its host 55 Cancri, which is a *G*-type star similar to the Sun. It is suggested that 55 Cancri *e* is a carbon-rich solid planet with a considerable proportion of it in the form of diamond! 55 Cancri *e* displays large variations in temperature across its surface (see Figure 6.2), possibly because of underlying volcanic activity that keeps emitting thick clouds of dust blocking thermal emissions from parts of the surface.

6.5 Ring Systems

We are familiar with the rings that exist around the distant, long-period gas planets in our solar system. In general, the planets have to be far away from the star's heat to retain icy rings around them. Although Saturn's icy rings are the most famous, rings do exist around the dusty Jupiter and the carbon-rich Uranus and Neptune. These rings are also referred to as planetary ring systems. Rings form around a planet when asteroids and comets pass too close to it. If these encounters are beyond the Roche limit or Roche radius, then the self-gravitating forces holding the components of the asteroid or comet sustain against the tidal forces tending to rip it apart. Within the Roche limit, which is approximately 2.5 times the radius of the planet, the asteroid or comet will disintegrate. As a general rule, rings are mostly icy around planets beyond the frost line, while rings consisting of rocky grains are more stable closer to the host star. As astronomers detect more and more exoplanets, it is possible they will come across exoplanetary ring systems also. Theoretically, as a ringed exoplanet moves in front of its host star, it will dim the star in a way entirely different from a non-ringed planet (Carlson et al. 2019).

In 2012, Eric Mamajek of Cerro Tololo Inter-American Observatory in Chile observed a mysterious object eclipsing its parent star in a very strange manner, unlike a spherical object or a circumstellar disk. Mamajek and his student, Mark Pecaut of Rochester University, noticed several dips in the light curve of the host star as the unique object orbited around it. The only plausible explanation was multiple thin dust rings orbiting an exoplanet. The star in question is 1SWASP J140747.93-394542.6 [variously known as 1SWASP J140747, J1407 and Mamajek's Object (see Figure 6.3)] at 420 light years away, which is a sun-like star but is much younger at 16 million years of age, being orbited by an object with large Saturn-like rings. Since the mass of 1SWASP is not known, it could be a brown dwarf or a low-mass star (Mamajek et al. 2012).

FIGURE 6.3 Scaled artistic representation of Mamajek's Object (Credit: Exoplanetary Science)

The solar system gas giants like Jupiter and Saturn have rings aligned with their equatorial plane. But exoplanets orbits are so close to their host stars that only the outermost rings of a planet are aligned with the planet's orbital plane around the star. However, the inner rings would be still aligned with the planet's equatorial plane. If the planet has a tilted rotational axis, then the varying alignments between the inner and outer rings would lead to a warped ring system (Schlichting et al. 2011).

6.6 Insolation Pattern

The pattern of insolation (i.e. exposure to radiation from the host star) on an extrasolar planet has profound implications on its climate and possible habitability.

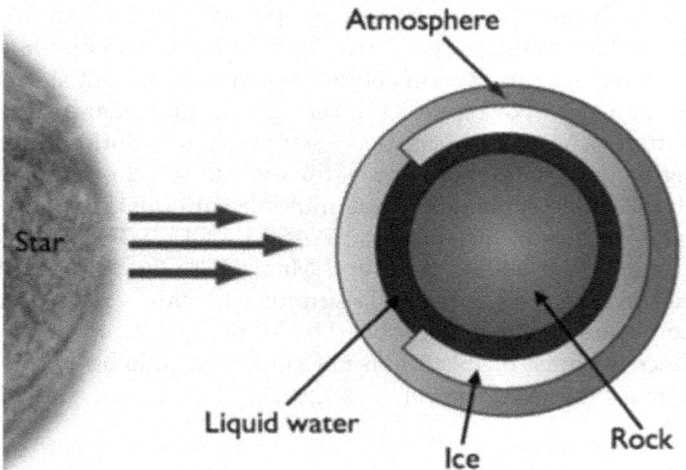

FIGURE 6.4 Schematic representation of an icy eyeball planet (Credit: Pierrehumbert 2011)

A planet's insolation regime depends on parameters like orbital eccentricity, obliquity of its spin axis, rotation rate, and longitude of the vernal equinox. For example, if a planet receives the same time-averaged insolation at both poles, the peak insolation at its poles can differ by a factor up to 27, depending on its eccentricity and equinox. This is important for planets with polar ice caps (or lakes and seas) like Mercury, Earth, and Mars (or Titan) (Dobrovolskis 2013). Dobrovolskis (2015) describes the insolation pattern of some planets as a striped ball, eyeball (Figure 6.4) or double-eyeball pattern, which ultimately changes the distribution of liquid and ice.

CHAPTER 7

Exoplanet Dynamics

"The chief aim of all investigations of the external world should be to discover the rational order and harmony which has been imposed on it by God and which He revealed to us in the language of mathematics."

– Johannes Kepler

7.1 Introduction to Kepler Laws

Kepler's laws of planetary orbits and dynamics play a key role in the understanding of the gravitational interaction between a star and its host planet (Murray and Correia 2010).

Box 5. *Johannes Kepler (1571-1630) was a renowned German astronomer and mathematician. He was an avid sky observer and arrived at three laws regarding the motion of planets around the Sun. He was aware of only six planets orbiting the sun: Mercury, Venus, Earth, Mars, Jupiter and Saturn. At the age of 27, Kepler started working as assistant to the Danish astronomer Tycho Brahe at the Royal Observatory in Prague. After 22 years of ceaseless work, Kepler arrived at three laws regarding the motion of planets around the Sun- the first two laws in 1609 and the third law in 1619. Kepler did not call his conclusions of planetary motion as laws. He simply regarded them as celestial harmonies that reflected God's design for the universe (Robert S. Westman). In 1627, Kepler published Tabulae Rudolphinae (The Rudolphine Tables) containing a wealth of immaculately accurate calculations of the positions of the solar planets worked out by Tycho Brahe. This monumental book stands as a testimony of the validity of Kepler's planetary laws. Around 50 years later, Sir Isaac Newton showed that Kepler's laws of planetary motion can be derived based on the Universal Law of Gravitation proposed by him.*

In astronomy, Kepler's three laws of planetary motion are useful in describing the motion of planets around the Sun or of any other celestial object under the action of a central inverse square force (Box 5). Let us look at each of these laws in detail without touching upon the mathematics.

I law of elliptical orbits: Planets move around the Sun in elliptical orbits with the Sun located at one focus (Figure 7.1). This law gives the shape of planetary orbit.

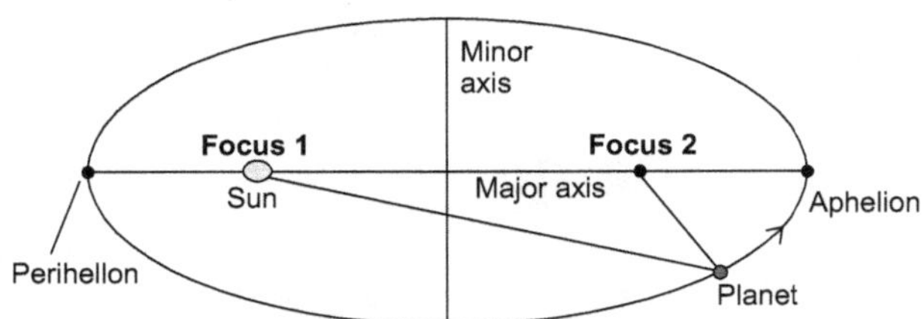

FIGURE 7.1 Elliptical orbit of a planet round the Sun

Kepler initially believed that a circle was the Universe's perfect shape and that planetary orbits must be circular in shape. His efforts to fit Tycho Brahe's detailed observations of the orbit of Mars into a circle were futile. Eventually, Kepler admitted that planetary orbits are stretched out circles, what we know now as ellipses. A circle is just a special case of an ellipse with major axis equal in length to minor axis and the two foci approaching each other and coinciding at the center of the circle. Take a torch and hold it perpendicular to a wall in a dark room. If you switch on the torch, you would see a circular patch of light on the wall. If you now hold the torch slightly inclined to the wall, the patch of light will change its shape into an ellipse.

An ellipse is a closed curve such that the sum of the distances of any point along it from Focus 1 and 2 is always a constant. In a circle, the two foci coincide at the center. The distance of any focus from the center can be expressed as a fraction 'e' of the semi major axis 'a'. 'e' is known as the eccentricity of the ellipse and is a measure of the extent to which its shape is deviated from that of a circle.

We know that a planet orbits the Sun under the mutual gravitational force of attraction, which obeys the inverse square law of distance-strength of the force varies inversely as the square of the distance between the centers of mass of the planet and the sun. Actually, both the planet and the Sun orbit around their barycenter or their common center of mass which is located deep within the Sun. While the planet being lighter goes round the Sun, the more massive Sun just wobbles about the barycenter. It is this wobbly motion of distant stars that has helped scientists detect exoplanets around them.

Kepler used the fact that under the action of a central force, angular momentum of a planet remains conserved, both in magnitude and direction and hence the elliptical orbit of the planet should be confined to a plane. Kepler mathematically arrived at the general differential equation in polar coordinates for the orbit of a mass moving under a central inverse square law force. The equation turned out to be identical with that of a conic section-the curve obtained by the intersection of a circular cone with a plane inclined at different angles as shown in Figure 7.2. It could be an ellipse with eccentricity $e < 1$, a parabola with $e = 1$, a hyperbola with $e > 1$ or a circle with $e = 0$.

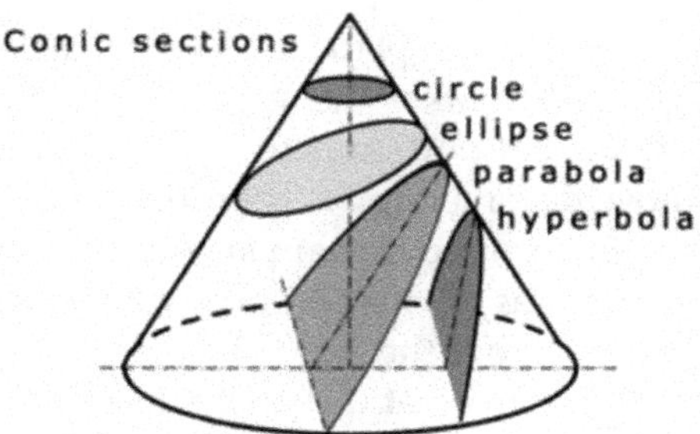

FIGURE 7.2 Conic sections (Credit: Magister Mathematicae)

Elliptical and circular orbits are closed and allow the planet to repetitively go round the object exerting a central force on it, similar to the motion of the moon round the Earth. While an ellipse is the most general orbit, a circle is just a special case. Parabola and hyperbola are open-ended and are not compatible with periodic orbits. A parabolic path with its branches becoming parallel to each other is a borderline case between closed and open orbits, and hence defines speeds in excess of or equal to escape velocity-minimum velocity required to escape from the gravitational pull of the Earth. A hyperbolic path with its divergent branches extending to infinity (Figure 7.3) is typical of interplanetary flights wherein the speed exceeds that along a parabolic path.

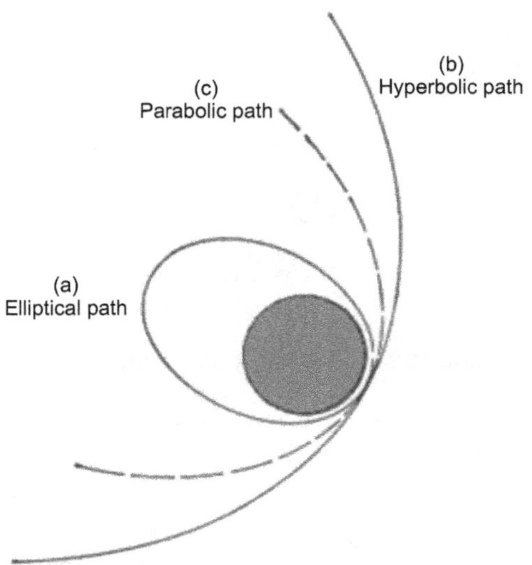

FIGURE 7.3 Conical paths (Credit: NASA)

II law of Areas: The radius vector joining the centers of mass of the sun and a planet sweeps out equal areas in equal intervals of time. In other words, the areal velocity of a planet is constant. This law gives the relation between the orbital speed of the planet and its distance from the Sun.

In Figure 7.4, the planet takes the same time to move along the arcs *AB*, *CD* and *EF*. The shaded portions are equal in area. This implies that the planet has to move faster along *BA* than along *FE*. In other words, the speed of a planet in orbit around the Sun keeps varying along its elliptical path. The speed increases as the planet approaches the Sun and decreases as it moves farther away from the Sun. The planet has maximum speed when it is closest to the Sun, i.e. at the perihelion and least speed when it is farthest at the aphelion. This is compatible with the law of conservation of mechanical energy. If m is the mass of a planet with an instantaneous velocity v, then its kinetic energy is $\frac{1}{2}mv^2$. If M is the mass of the Sun and r is the instantaneous distance between the Sun and the planet, then its potential energy is $-GMm/r$, where G is the gravitational constant and the negative sign indicates that central force is attractive in nature or work has to be done (energy has to be expended) to increase the distance between the Sun and the planet.

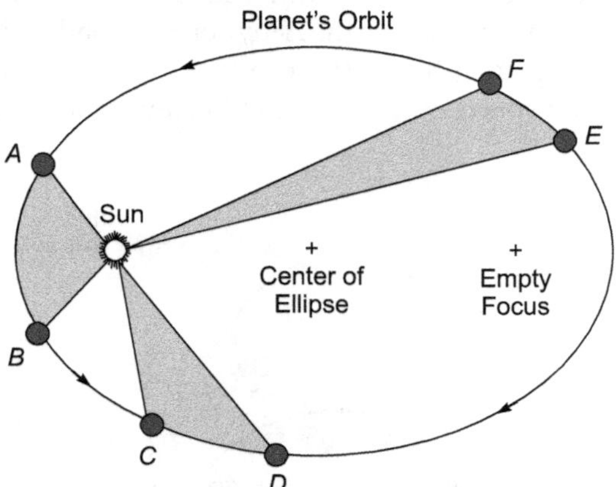

FIGURE 7.4 Kepler's second law of planetary motion (Credit: NASA)

Total mechanical energy of the planet is given by

$$E = \frac{1}{2}mv^2 - \frac{GMm}{r} \tag{7.1}$$

As r increases, v decreases, i.e. as a planet moves away from the Sun, its speed decreases and vice versa.

III Law or Harmonic Law: Square of the orbital period of a planet or planet's year 'T' is directly proportional to the cube of the semi major axis 'a' of its elliptical orbit, i.e. $T^2 \propto a^3$. This law gives the relation between the size of the orbit of a planet and the period of its revolution around the sun. This also implies that all planets in elliptical orbits with the same major axis have the same period of rotation round the Sun. It is possible to determine the distance of a planet from the Sun knowing its orbital period and vice versa.

7.2 Dynamics of Keplerian Orbits

Discovery of over 4000 extrasolar planets till date has affirmed the possibility of a wide range for their orbital parameters and physical properties. The host stars generally belong to the F to M spectral types. The exoplanets around these stars vary in distance from the host stars from 0.014 AU to 670 AU, can range in mass from 2 to 20 times that of Jupiter, with orbital periods ranging from 1 day to more than 15 Earth years and orbital eccentricities ranging from perfect circular orbits ($e = 0$) to most elliptical with $e > 0.9$ (Murray and Correia 2010).

Understanding the Kepler problem or two-body problem is challenging in the case of exoplanets.

Newton demonstrated that the elliptical path arises directly by solving the basic universal law of gravitation force proposed by him:

$$F = -\frac{Gm_1m_2}{r^2} \tag{7.2}$$

where F is the gravitational force of attraction between two masses m_1 and m_2, which are the masses separated by a distance r and G is universal gravitational constant.

But finding the instantaneous position and velocity of an object is difficult in the two-body problem. The solution to this was given by Kepler using Newton's derivation, to find an elliptical path from solving the basic universal law of gravitation. (Note: The entire derivation is available in Murray and Correia 2010).

Although Kepler's laws describe the motion of celestial objects accurately over short spans on an astronomical scale, they are only approximate in the true sense. The laws are strictly applicable to a two-body problem in which an object orbits another large fixed object of spherical mass distribution under the action of a single central force. The simplest instance of the earth going around the Sun is strictly not a two-body problem; firstly because the Sun is not a stationary anchor of the solar system. As every solar planet gravitationally tugs on the

Sun, the Sun ends up orbiting round the barycenter, the center of mass of the solar system. Further, barycenter by itself is not a fixed point but keeps varying depending on the instantaneous relative positions of the planets around the Sun. Added to this is the mutual force of gravitational attraction among the planets. Neither the Sun nor any of the solar planets are perfectly spherical with uniform mass distribution. Because of their spin motion, the sun and the planets are oblate spheroids, slightly bulging round the equator and flattened around the poles. All these factors perturb the parameters of the Keplerian elliptical paths of planets around the Sun, and the resulting perturbed elliptical orbits are referred to as Osculating orbits. However, these perturbations are negligible since planetary masses are miniscule compared to the Sun's mass and hence Kepler's laws describe the motion of solar planets accurately.

In actuality, the solar system or any other exosolar star system with its exoplanets is an n-body problem which has no complete closed solution.

Also, the gravitational central force is basically conservative in nature, independent of the velocities of the bodies involved. In other words, there is no dissipation of mechanical energy due to friction and other causes. However, if the orbiting body is very closely bound to the central body, it may experience a tidal distortion in its shape. Energy gets dissipated in this tidal coupling.

7.3 Non-Keplerian Dynamics of Exoplanets

Extreme motions of exoplanets with short periods or high eccentricities can be understood only through a different equation of motion of the two-body problem taking into account non-Keplerian effects. The equation for the instantaneous acceleration *'a'* of a planet of mass *'mp'* when it is at a distance of *'r'* from the host star of mass m^* is given by

$$a = -G(m^* + m_p)\, r/r^3 + f \tag{7.3}$$

where G is the gravitational constant and *'f'* is a force other than the mutual force of gravity between the planet and the host star. $f = 0$ implies Keplerian motion.

This can be further extended to incorporate relativistic effects (Fabrycky and Tremaine 2007). With a deeper understanding of the dynamical calculations it is possible to detect small exoplanets, which forms the baseline for a new technique of detecting exoplanets, known as the Transit-timing method. Through a study of non-Keplerian motion of exoplanets, it is possible to understand the history of their dynamical configuration.

CHAPTER 8

Exoplanet Formation

"We had the sky, up there, all speckled with stars, and we used to lay on our backs and look up at them, and discuss about whether it was made, or only just happened"

– Mark Twain, Huckleberry Finn

8.1 Protoplanetary Disks and Debris Disks

Much of the space between stars, i.e. the interstellar space, consists of gases, mostly hydrogen and helium in atomic and molecular forms, along with dust or solid particles of mainly carbon, silicon and oxygen, and some plasma. This initially tenuous combination of gas and dust becomes a huge interstellar cloud if it gets denser than the mean density of the interstellar medium (ISM). Most of these self-gravitating, cool clouds collapse naturally. At times, shock waves due to supernova explosions of nearby dying stars can strike the gas nebula and accelerate the process of collapse. As the cloud collapses, its core temperature and density keep increasing. When the core is hot enough for its pressure to balance the inward gravitational pressure, collapse of the nebula stops and at this stage, the celestial object is referred to as a protostar. The rotating core starts spinning rapidly when it reaches the size of a star. This rotation sustains a disk of gas and dust around it in the form of a disk, called the protoplanetary disk. On reaching hydrostatic equilibrium, the star stops contracting and is recognized as a main sequence star. The star becomes luminous through fusion burning of hydrogen to helium. This results in the surrounding material being blown away. Protoplanetary disks evolve over timescales of millions of years and undergo significant mass loss mainly due to photoevaporation (Wyatt 2008). What remains over decides whether planets can form around the host star or not. Collisions with planetesimals in orbit around the star results in the accretion of a second generation of gas and dust around the star in the form of a debris disk or secondary disk. Physicists of the 17th century suggested that the solar system planets were formed out of a rotating debris disk surrounding the young Sun.

Protoplanetary and debris disks are different with regard to their composition and structure. While the former are optically dense, the latter are optically thin across the electromagnetic spectrum. Both are identified by their infrared emission and fractional luminosity. Primordial hydrogen is prominently absent in debris disks.

During the earlier timeline, it was surmised that the gaseous envelope would contract under its own gravity and conservation of angular momentum would flatten it out into a disk. This stage is referred to as class '0' of the star being formed. This explains the solar planets' coplanar and nearly circular orbits around the Sun. More than two and a half decades ago, through a detailed observation and analysis of their spectra, it was proved that the gas and dust disks around young stars are in a state of evolution (Weinberger et al. 1999, Kawabe et al. 1993).

Currently available techniques for observation of these gas disks allow us to understand the process of planetary formation in detail. Also, the discovery of exoplanets has revealed that they can exist in a wide variety of environments. Both these concepts are helpful in a better understanding of the characteristics of our own solar system (Isaacman and Sagan 1977).

The evolution of a turbulent planetary disk and the formation of a planetary system are closely linked to the energetic UV and X-radiation from the parent star and also the interstellar radiation field.

8.2 Terrestrial and Giant Planet Formation

Till the first exoplanet was discovered in 1995, our understanding of the formation of planets was restricted to those belonging to the solar system. With over 4000 exoplanets being cataloged as of date, and given their diverse properties like mass, size, density, proximity to host stars and orbital eccentricity, it is necessary to explore their evolution in a detailed manner (Helled and Vazan 2013).

Our solar system is more organized, with a clear demarcation between the domains of inner rocky terrestrial planets and the outer gas giants by the presence of the asteroid belt between Mars and Jupiter. Traditionally, it is believed that the solar planets formed in the whereabouts of their current distance from the Sun and have remained there ever since. Terrestrial planets form through a process known as Core Accretion (CA). During CA, planetesimals ranging from a few tens of meters to a few hundred kilometers build up on an embryo through the coagulation of microscopic particles which stick together electrostatically. The process continues till a heavy, molten metallic (referring to elements other than hydrogen and helium) core of the planet is formed. Most lighter elements like hydrogen and helium get swept away by the solar wind (Redd 2016). While the denser elements sink to the core of the planet, the lighter elements form the mantle and the crust. Collisions during plate tectonics result in the formation of physical features like mountains, valleys and volcanoes.

Gravitationally trapped gases form a blanket of atmosphere around the planet. Core accretion is inadequate to account for the formation of gas giants, which hold on to enormous gases in them.

8.3 Planetary Migration

In the past two decades, nearly 4000 exoplanets have been discovered and confirmed. Many of these exoplanets have bizarre orbits around their host stars and may not have formed where they are seen today. In other words, the exoplanetary orbits are in a dynamic state of evolution, which is referred to as *planetary migration*. Migration refers to the physical movement of a planet over time with respect to its host star, resulting in a change in the size of its orbit and hence position. Figure 8.1 indicates the steps involved in tracing the migration history of a planet.

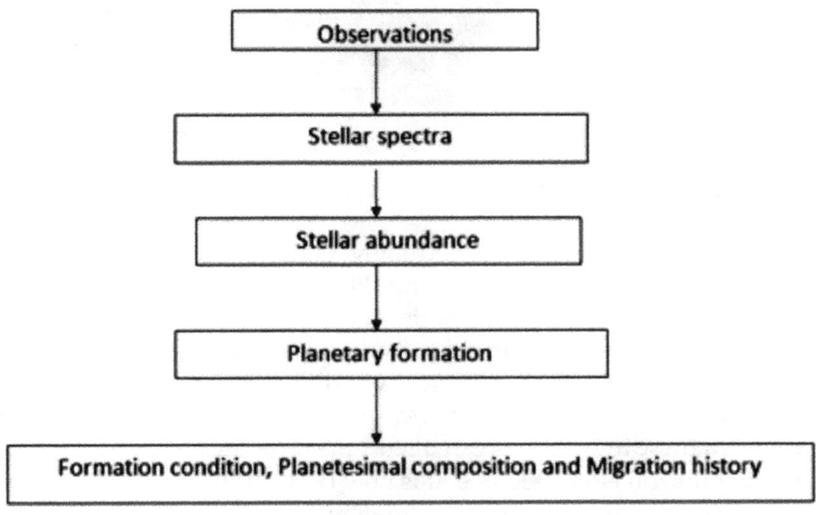

FIGURE 8.1 Schematic tracing of migration history of a planet

As per the nebular hypothesis, exoplanets discovered around young stars form from the protoplanetary disk of dust and gas. Once formed, the orbit of a planet may change due to interaction with gas disks, other planets, planetesimals and small objects in the system (Armitage 2008). This is referred to as planetary migration. Migration of planets is understood to be of three types: I, II and III, each influenced by a different type of dynamical interactions of the planet with its environment.

Type I migration is applicable to planets with low masses. Typically, planets of mass upto ten Earth-mass migrate towards the host star without affecting the basic structure of the protoplanetary disk, also called the embryonic disk. Due to the weak gravitational interaction with the host star, the planet initiates a spiral wave in the gaseous disk, robbing it of its angular momentum (Refer Figure 8.2 for all migration types). This drives the orbit of the planet to shrink and the planet slowly moves towards the host star till it settles down in a stable orbit. All this happens over time scales of a few million years, without significantly affecting the surface density profile of the gas disk (Armitage 2008).

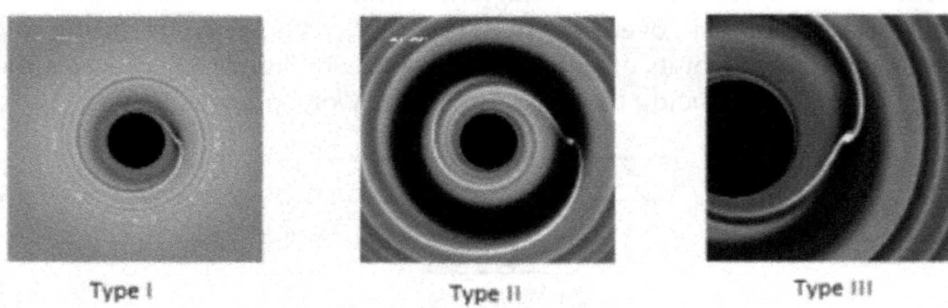

FIGURE 8.2 Planetary migration types (Credit: Frédéric Masse)

More massive planets that strongly perturb the gas disk undergo *Type II* migration. The exchange of angular momentum between the planet and the host star results in the repulsion of the gas from the vicinity of the planet. This results in an annular gap around the planetary orbit. The planet can keep accreting some gas via thin streams crossing the gap. Accretion rate decreases as the planet grows in size and the gap becomes deeper. All these interactions can distort the near circular original orbit of the planet into an elongated one. It is now believed that Type II migration accounts for the existence of Hot Jupiters of short periods in eccentric orbits.

Type III migration, also referred to as 'runaway migration', is characterized by extremely short time-scales of migration of planets through partially open gaps in the gas disk. The planet's initial rapid rotational motion displaces gas from the disk in co-orbital regions. This creates an imbalance in the density of gas in the leading and trailing side of the planet. This drives the planet inwards to an orbit closer to the host star.

CHAPTER 9

Exoplanet Interiors and Atmosphere

"There are often evolutionary parallels on the different worlds because creation tends to be economical."

– Julian May

9.1 Terrestrial and Giant Planet Interiors

As per planetary scientists, the dynamics of a planet's interior are essential for understanding its habitability (Kaufman 2019). Solar planets are conveniently classified into terrestrial planets and gas giants. Mercury, Venus, Earth and Mars, the four inner planets of the solar system, are called terrestrial or telluric planets, while Saturn, Uranus, Neptune and Jupiter are referred to as gas giants. What distinguishes the two categories is the ability of a planet to retain an atmosphere predominantly of primordial hydrogen and helium around it (Sotin et al. 2010) depending on its mass. It is proposed that a planet less than 15 times the mass of Earth will lack primordial H and He in its atmosphere (Wuchterl et al. 2000) and therefore will be terrestrial in nature. Planetary habitability strongly depends on how its atmosphere is linked to and configured by its internal structure.

Terrestrial planets of the solar system are also referred to as rocky planets since their central, dense cores consist primarily of rocks and metals. Most abundant rocks in the core are silicates, made of silicon and oxygen, and the most common metal in the core is iron. Venus, Earth and Mars have almost similar silicate dominated composition: two-thirds of silicates, remaining one-third of iron-nickel or iron-sulfur rocks. Mercury has the largest component of metals, iron in particular and our Moon has the least. Within the core, the denser metals are towards the center, while the lighter silicates are towards the outer edge. This suggests that although terrestrial planets are solid masses today, at some epoch of formation they must have been hot enough to have melted. Cores of terrestrial planets have a sheath of solid mantle of lower density, with an outermost solid shell called the crust. Easily detectable giant gas planets like Jupiter and Saturn, also referred to as Jovian planets, are mostly composed of primordial hydrogen and helium, and lack solid or liquid surfaces essential for supporting life. Temperature rises rapidly as we delve deeper into the interior of gas giants-too hot to support any form of life. Jupiter and Saturn have a similar composition: a dense core of rocks and ice, mostly of liquid metallic hydrogen (Howell et al. 2018), surrounded by a gaseous atmosphere of mostly hydrogen, helium, ammonia and methane. Ice giants like Uranus and Neptune have a thinner atmosphere of hydrogen, helium and methane enveloping a mantle of water, ammonia and methane ice, with a central rocky and icy core.

Figure 9.1 schematically represents the steps involved in arriving at the internal composition of terrestrial planets and gas giants.

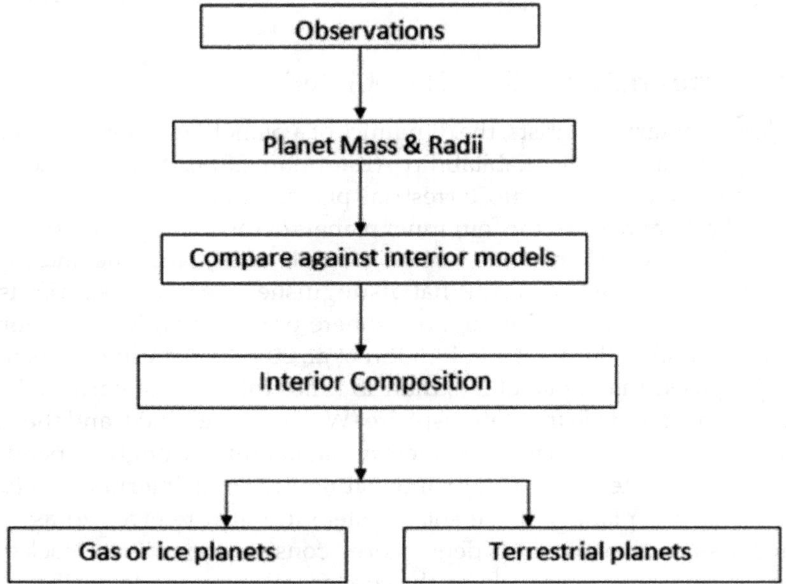

FIGURE 9.1 Schematic description of interior composition of terrestrial planets and ice giants

A habitable exoplanet is characterized by:

(i) A stable temperature supported by the host star or some geological activity (Storbel astronomy notes 1998).
(ii) Liquid water to enable biochemical reactions.
(iii) The presence of carbon, hydrogen, nitrogen, oxygen, phosphorus and sulfur, and transition elements like iron, chromium and nickel which are the essential building block elements.
(iv) Telluric with a solid surface.
(v) Sufficient mass to gravitationally retain an atmosphere around it.

(vi) Plate tectonics to regulate the surface temperature via the carbon cycle, to cool the interiors fast enough to generate a magnetic field to shield the planet from stellar winds, harmful radiation and water loss, to create a dry land on an otherwise water covered planet and to nurture a high level of biodiversity on the planet.

(vii) A relatively large moon close by to prevent the rotational axis of the planet from tilting too much or too quickly.

9.2 Terrestrial and Giant Planet Atmospheres

Search for life on terrestrial exoplanets is one of the most significant and exciting fields of research (Sotin et al. 2010). The modeling of exoplanetary atmosphere mainly depends on two factors: a) physical conditions of gas giants, and b) super Earths that are transiting young gas giants that have been detected by direct imaging. Other ways of modeling are:

1. Computational complexity (i.e. from one-dimensional (1-D) plane-parallel models to three-dimensional (3-D) general circulation models (GCMs)),
2. Thermochemical conditions (including models assuming solar-composition in thermochemical equilibrium as well as those with non-solar abundances and non-equilibrium compositions).

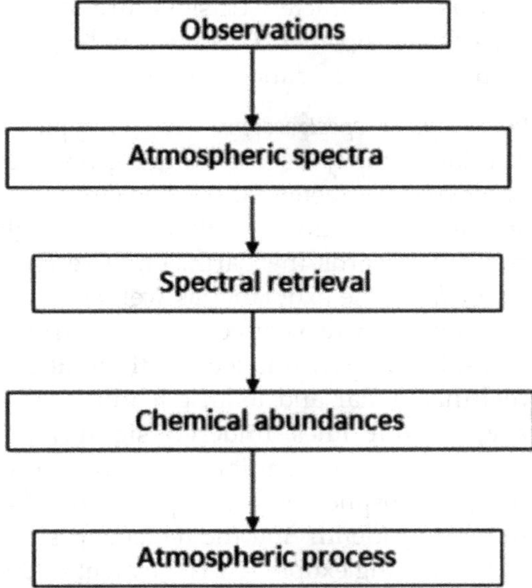

FIGURE 9.2 Schematic diagram of atmospheric process in a planet

Planetary atmosphere holds many key facts regarding the various physical and chemical processes in action, which in turn provide information regarding the evolutionary history of the planet. The study of exoplanetary atmospheres has progressed at a tremendous pace in recent years (Madhusudhan 2019). A simple flowchart to arrive at the atmospheric process of a planet from observational techniques is shown in Figure 9.2. Many exoplanets have been observed to have a gravitationally bound atmosphere around them. It is easier to detect the extended atmospheres of Hot Jupiters which orbit close to their host stars. There are several methods for obtaining spectra of exoplanetary atmospheres. These spectroscopic methods depend on the transmission, reflectance and thermal emission of the atmosphere. The chief techniques employed by both from space as well as the ground-based facilities are:

(i) Transit spectroscopy: This is an interesting option when the exoplanet transits across the host star and a small fraction of the star light is transmitted through the tenuous atmosphere of the exoplanet. This method allows the planetary atmosphere to be observed in three configurations: transition spectrum when the planet transits in front of the host star, i.e. at primary eclipse, emission spectrum while the planet moves behind the star, i.e. at secondary eclipse, and the spectrum consisting of a continuous spectrum due to the host star in the background with a line absorption spectrum in the foreground, similar to Fraunhofer lines in the solar spectrum. Each line is the characteristic signature of the element present in the atmosphere of the exoplanet. This method helps in arriving at the thickness of the atmosphere as a function of wavelength.

(ii) High resolution Doppler spectroscopy: This is typically a reflectance spectroscopy technique widely used by large ground based telescopes for exoplanet atmosphere characterization. The method uses the fact that when an exoplanet is approaching occultation, some light from the host star may bounce off it towards the Earth. Light received on Earth is a combination of light from the exoplanet as well as direct light from the parent star and the spectrum also has contributions from telluric features of the Earth's atmosphere. Spectra taken by different observatories over many years reveal that stellar and telluric features remain unchanged, and exoplanetary spectral lines undergo significant Doppler shifts (Hoeijmakers et al. 2019). This technique is successfully used to detect water and CO in the atmosphere of Hot Jupiters (Brogi et al. 2012). This technique is also used in identifying the molecular signatures from the atmosphere of non-transiting exoplanets (Watson et al. 2019).

(iii) Emission spectroscopy: This method is employed to detect atmospheres of exoplanets which are orbiting close to their host stars. The planet itself may be hot enough due to its internal heat sources or may be irradiated sufficiently by the host star to emit radiation similar to a blackbody. This low resolution direct imaging technique has helped in the discovery of fewer exoplanetary atmospheres as compared to the other two methods (Madhusudhan 2019).

Planetary atmospheres can be classified into "primary" or "secondary" atmospheres. Primary atmospheres are accreted directly from the nebula, which is controlled by hydrogen and helium gas. Secondary atmospheres include volatiles outgassed from the planet's interior. In our solar system, Jupiter, Saturn, Uranus, and Neptune possess primary atmospheres, while rocky planets and moons have secondary atmospheres. Uranus and Neptune atmospheres are only 10-20% hydrogen and helium by mass suggestive of some minor secondary components too. Generally, planets in the mass range of 1 to 15 solar mass have varied atmospheres that could either be fully primary or fully secondary or a mixture of both. The accretion of primary atmospheres is modeled by the core accretion theory of planet formation (Helled et al. 2014). The outgassing of secondary atmospheres is also modeled for the solar system, but not much work has progressed for exoplanets (Elkins-Tanton and Seager 2008, Rogers and Seager 2010, Madhusudhan 2019).

9.3 Atmospheric Circulation

Atmospheric circulation refers to a large-scale motion of the air, and oceanic circulation, which helps in the understanding of the thermal energy distribution on the surface of the planet. From Showman et al. (2013), the atmospheric circulation of planets that rotate more slowly or have a thicker atmosphere allows more heat to flow to the poles which reduces the temperature differences between the poles and the equator.

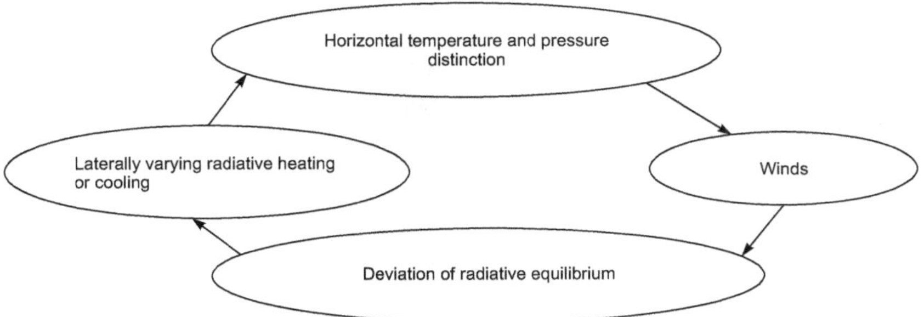

FIGURE 9.3 Schematic diagram of atmospheric circulation

Let us have a look at the factors that impact atmospheric circulation.

Horizontal temperature variation implies the presence of horizontal pressure variation, which in turn drives winds. The winds push the atmosphere away from the radiative equilibrium zone by transporting heat from hot regions to cooler regions (e.g. in our Earth from the equator to the poles). This temperature and pressure variation from radiative equilibrium gives net radiative heating and cooling to occur, which maintains the horizontal temperature and pressure contrasts that drive the winds (Figure 9.3). Spatial variation in thermodynamic heating/cooling drives the circulation, yet it is the presence of the circulation which allows these heating/cooling patterns to exist. The atmospheric circulation now becomes a coupled radiation-hydrodynamics problem. For example, on Earth, the equator and poles are not in radiative equilibrium. The equator is conditioned to total net heating and the poles to total net cooling- this is the mean latitudinal heat transport that is both responsible for and driven by these net imbalances. The mean climate (for example, the global-mean surface temperature of a planet) depends on the absorbed stellar flux and the atmosphere's need to reradiate that energy to space. Here the global-mean climate is affected by the atmospheric mass, composition, and circulation. On Earth-like terrestrial planets, the circulation controls the distribution of clouds and surface ice, which determine the planetary albedo and the mean surface temperature. For some cases, a planetary climate can have multiple equilibrium (for example: a warm, ice-free state or a cold, ice-covered "snowball Earth" state), and in such scenarios the circulation plays an important role in determining the relative stability of these equilibrium. Understanding the atmosphere/climate system is important and challenging due to its nonlinearity, which involves multiple positive and negative feedbacks between radiation, clouds, dynamics, surface processes, planetary interior, and life (if it exists).

CHAPTER 10

Mathematical Indexing Formulations of Exoplanets

"How can it be that mathematics, being after all a product of human thought which is independent of experience, is so admirably appropriate to the objects of reality?"

– Albert Einstein

Parameters such as mass, radius, density, surface temperature and escape velocity of exoplanets can be obtained by employing any one of the detection techniques. Knowing these parameters, it is possible to understand how similar the exoplanets are to Earth and Mars and their potential habitability.

10.1 Earth Similarity Index

Earth similarity index (ESI) or the "easy scale" is a multiparameter characterization of an exoplanet to estimate how similar it is to the Earth. ESI is arrived at by computing the radius, density, effective temperature and escape velocity of an exoplanet. ESI has a scale range from zero to 1 with Earth having a value of 1, 0 being dissimilar and 1 being similar to Earth. ESI values above 0.8 may be considered 'Earth-like' (Schulze-Makuch et al. 2011). For example, exoplanet Kepler-438b discovered in 2015 by NASA's Kepler space telescope has the highest ESI of 0.83 for any exoplanet and it is the first potentially Earth-like planet orbiting in the habitable zone of a Sun-like host star.

According to Kashyap et al. (2017), distance/similarity measurements are widely used in the classification of objects in various disciplines (Deza and Deza 2006). Here, the distance *'d'* is represented as dissimilarity and proximity is taken as equivalent to similarity *'s'*. Mathematically, the concept of distance is a metric one and is a measure of true distance in Euclidean space Rn. This problem is usually addressed using Minkowski's space of L_p form (Cha 2007) in which p-norm stands for finite n-dimensional vector space:

$$d = (\Sigma_{i=1}^{n} |p_i - q_i|^{p})^{1/p} \tag{10.1}$$

where the Manhattan L_1 distance, $d_{\text{Man}} = \Sigma_{i=1}^{n} |p_i - q_i|$ (10.2)

and Euclidean distance L_2, $d_{\text{Euc}} = (\Sigma_{i=1}^{n} (p_i - q_i))^{1/2}$ (10.3) are the special cases. Here, p_i and q_i are the coordinates of p and q with dimension $i = 1, 2, 3\ldots, n$. L_1 form has an advantage of being decomposed into contributions made by each variable as the sum of absolute differences. For example, for the L_2 form, it would be the decomposition of the squared distances. Our interest is to find similarities between different planets based on their various characteristics. In such cases, the Bray-Curtis distance is the most widely used scale (Bray and Curtis 1957, Greenacre and Primicerio 2013). Bray-Curtis is a modified Manhattan distance, where the summed differences between the variables are standardized by the summed variables of the objects:

$$d_{BC} = \frac{\sum_{i=1}^{n} |p_i - q_i|}{\sum_{i=1}^{n} (p_i + q_i)} \tag{10.4}$$

Here p_i and q_i are two different precisely measurable quantities between which the distance is to be measured and n is the total number of variables. The Bray-Curtis scale assumes that the samples are taken from the same physical measure, say, mass or volume. Since the distance is computed from the raw counts, a higher abundance in one sample as compared to another becomes a part of the difference between the two samples. In Bray-Curtis scale, the interpretation is such that zero means the samples are exactly the same and one means they are completely dissimilar. It should be recalled that Bray-Curtis distance is not the true metric distance but is a semi-metric distance (in which the distance between two distinct points can be zero), which is usually called dissimilarity, or ecological distance (Kindt and Coe 2005). The advantages in using the ecological distance is that differences between datasets can be expressed by a single statistic.

The intersection between two distributions is a more widely used form of similarity (Looman and Campbell 1960). Most similarity measures for intersection can be transformed from the distance measure using the transformation technique but not exclusively (Bloom 1981),

$$S_{BC} = 1 - d_{BC} = 1 - \frac{\sum_{i=1}^{n} |p_i - q_i|}{\sum_{i=1}^{n} (p_i + q_i)} \tag{10.5}$$

Here, the value of 0 means complete absence of relationships and the value of 1 shows a complete matching of the two data records in the n-dimensional space (Schulz 2007).

Distances/similarities based on heterogeneous data can be found after a process of standardization – balancing the contribution of different types of variables in an equitable way (Greenacre and Primicerio 2013). This is done by calculating the similarity for each set of homogeneous variables and then combining them using various methods, which are described in the following paragraphs while demonstrating the ESI calculation. Higher values in one set may influence the result of the Bray-Curtis similarity more dominantly and imply these variables are more likely to discriminate between sets. Therefore, user-defined weighting is a convenient (though subjective) method for down-weighing the differences for a set of variables. In the present work, our aim is to compare sets of different variables of one planet with that of the reference value, for example the Earth, to identify planets that are similar to Earth. Rewriting Eq. (10.5),

$$S = [1 - |X - X_0|/(X + X_0)]^{W_X} \tag{10.6}$$

where X is the physical property of the exoplanet, X_0 is the reference value, W_x is the weight for this property and the dimension $n = 1$, since we are constructing the index separately for each physical property. We find the weights by defining the threshold value (V) in the similarity scale for each quantity,

$$V = [1 - |X - X_0|/(X + X_0)]^{W_X} \tag{10.7}$$

In the literature by Bloom (1981), the similarity index ranging from zero to one is subdivided into 0.2 equal intervals, defining very low, low, moderate, high and very high similarity regions. The threshold is defined on the basis of intervals, for example considering only a very high similarity region with threshold value equal to $V = 0.8$. Fixing the threshold value V and defining the physical limits x_a and x_b as the permissible variation of a variable with respect to x_0 (i.e., $x_a < x_0 < x_b$), we can calculate the weight exponents for the lower W_a and W_b upper limits:

$$\begin{aligned} W_a &= \text{In}\,\text{V}/\text{In}\,[1 - |X_0 - X_a|/(X_0 + X_a)] \\ W_b &= \text{In}\,\text{V}/\text{In}\,[1 - |X_b - X_0|/(X_b + X_0)] \end{aligned} \tag{10.8}$$

The average weight W_X is obtained by the geometric mean of lower W_a and upper W_b limits,

$$W_X = \sqrt{(W_a \times W_b)} \tag{10.9}$$

For the present studies, the Earth and Mars similarity indices are defined as:

$$ESI_X = [1 - |X - X_0/(X + X_0)|]^{W_X} \tag{10.10}$$

$$MSI_X = [1 - |X - X_0/(X + X_0)|]^{W_X} \tag{10.11}$$

where X is the physical parameter of the exoplanet such as radius or density and X_0 is the reference to Earth for ESI and to Mars for MSI. The mean radius (R), bulk density (ρ), escape velocity (V_e) and surface temperature (T_s) of exoplanets are used as input parameters for computing similarity indices.

These parameters, except the surface temperature (which is expressed in kelvin), are used in Earth Units (EU) for the calculation of ESI and in the Mars Units (MU) for the calculation of the MSI. The corresponding weight exponents for both ESI and MSI scales were computed adopting the threshold value $V = 0.8$, indicating a very high similarity region. The weight exponents for the upper and lower limit of parameters were calculated for the Earth-like parameters (Schulze-Makuch et al. 2011) and presented in Table 10.1: radius range from 0.5 to 1.9 EU, mass range from 0.1 to 10 EU, density range from 0.7 to 1.5 EU, surface temperature range from 273 to 323 K and escape velocity range from 0.4 to 1.4 EU.

TABLE 10.1 Weight exponents of ESI

Planetary property	Reference values for ESI	Weight exponents for ESI
Mean radius (R)	1 EU	0.57
Bulk density (ρ)	1 EU	1.07
Escape velocity (V_e)	1 EU	0.70
Surface temperature (T_s)	288 K	5.58

In order to determine ESI of exoplanets, we converted all the input parameters to Earth Units (EU), except the surface temperature expressed in Kelvin. The corresponding ESI for each parameter (ESI_R, ESI_ρ, ESI_{T_s} and ESI_{V_e}) of the planet was calculated using Eq. (10.10). Using these ESI values, the interior ESI and surface ESI are determined using the following relations:

$$ESI_I = \sqrt{ESI_R \times ESI_\rho} \tag{10.12}$$

$$ESI_S = \sqrt{ESI_T \times ESI_{V_e}} \tag{10.13}$$

where ESI_R, ESI_r, ESI_{T_S} and ESI_{V_e} are ESI values calculated for radius, density, surface temperature and escape velocity, respectively. Then, the global ESI is calculated using the relation,

$$ESI = \sqrt{ESI_I \times ESI_S} \tag{10.14}$$

Sample values of ESI we calculated for two exoplanets are given in Table 10.2.

TABLE 10.2 Sample representation of ESI

Names	Radius (EU)	Density (EU)	Surface temperature (K)	Escape velocity (EU)	ESI_I	ESI_S	ESI
Earth	1.00	1.00	288	1.00	1.00	1.00	1.00
Mars	0.53	0.73	240	0.45	0.82	0.65	0.73
GJ 667Cc	1.54	1.05	288	1.57	0.92	0.91	0.92
Kepler-296 e	1.48	1.03	312	1.50	0.93	0.83	0.88

EU = Earth Units, where Earth's radius is 6371 km, density is 5.51 g/cm^3 and escape velocity is 11.19 km/s.

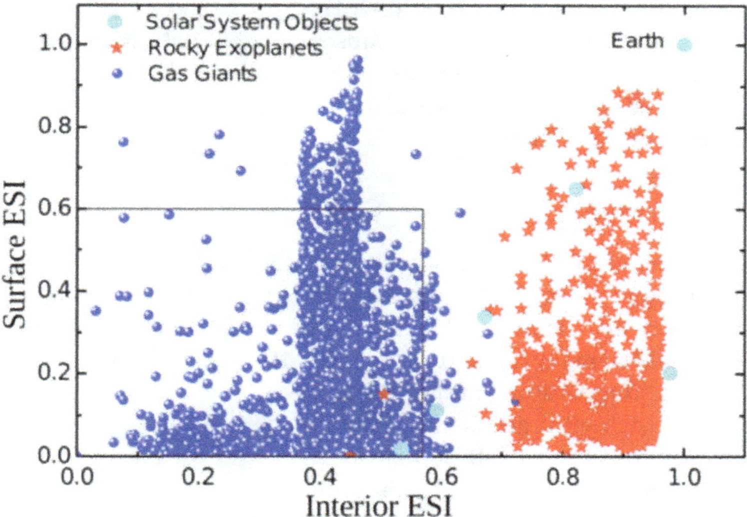

FIGURE 10.1 Interior ESI versus Surface ESI plot (Refer Kashyap et al. (2017) for an upgraded version)

Figure 10.1 scatter plot shows the ESI analysis from Kashyap et al. (2017). We found 29 Earth-like exoplanets. But this number keeps changing every year with new discoveries of exoplanets.

10.2 Planetary Habitability Index

Rory Barnes and his research associates at the Washington University's Virtual Planetary Laboratory suggested the use of a new Planetary

Habitability Index (PHI) in order to understand the habitability of exoplanets with varying chemical composition (Schulze-Makuch et al. 2011). PHI approach requires the presence of a stable substrate with appropriate chemistry to hold a liquid solvent capable of supporting life. But identifying the presence of a stable substrate having these features on all exoplanets is a challenging task.

$$PHI = (S\ E\ C\ L)^{1/4} \tag{10.15}$$

where S is the presence of a stable substrate, E is the available energy, C is the appropriate chemistry, and L is a liquid solvent in the world of interest. Each value is further subdivided into measurements of different physical quantities such as an atmosphere, a magnetosphere, quantity of heat and light received. The obtained value is finally normalized to the largest one (Schulze-Makuch et al. 2011). This reduces PHI to a fraction between 0 and 1 corresponding to 'absence of habitability potential' and 'potentially habitable' (Mozos and Moya 2017). PHI is another metric scale with geometric mean formulation that can be used to examine the habitability of exoplanets.

10.3 Mars Similarity Index

Mars Similarity Index (MSI) is defined similar to ESI but with respect to Mars-like exoplanets. The weight exponents for the lower and upper limit of parameters are defined for the Mars-like conditions: radius range from 0.72 to 1.88 MU, mass range 0.514 to 9.30 MU, density range 0.89 to 1.402 MU, surface temperature range 233 to 418 K and escape velocity range 0.85 to 2.23 MU with radius as 3390 km, density as 3.93 g/cm^3, the mean surface temperature as 240 K (Barlow 2014) and escape velocity as 5.03 km/s. The rationale behind the definitions of limits is to have a rocky planet with a lower limit in comparison to Mars (mass and radius are chosen as for Mercury, the smallest planet in our solar system) and with Earth as the upper limit. The temperature range is chosen based on the temperature known to be suitable for extremophile life forms, i.e. 40 to +145°C (Tung et al. 2005). The corresponding weight exponents for MSI were computed using the same method as for the ESI. Prior to calculating MSI of exoplanets, all the input parameters are converted to Mars Units (MU) except the surface temperature. MSI for each parameter (MSI_R, MSI_r, MSI_{T_s} and MSI_{V_e}) of the planet is calculated using Eq. (10.11). Using these values, the interior MSI, surface MSI and global MSI are calculated using the relations below:

Interior MSI; $$\mathrm{MSI}_I = \sqrt{\mathrm{MSI}_R \times \mathrm{MSI}_\rho} \quad (10.16)$$

Surface MSI; $$\mathrm{MSI}_S = \sqrt{\mathrm{MSI}_T \times \mathrm{MSI}_{V_e}} \quad (10.17)$$

Global MSI; $$\mathrm{MSI} = \sqrt{\mathrm{MSI}_I \times \mathrm{MSI}_S} \quad (10.18)$$

Here, MSI_R, MSI_r, MSI_{T_s} and MSI_{V_e} correspond to radius, density, surface temperature and escape velocity, respectively. Sample calculation of MSI for Earth is shown below along with tabulated input parameters:

TABLE 10.3 Weight exponents of MSI

Planetary property	Reference values	Weight exponents
Mean radius (R)	1 MU	0.86
Bulk density (r)	1 MU	2.10
Escape velocity (V_e)	1 MU	1.09
Surface temperature (T_s)	240 K	3.23

TABLE 10.4 Sample representation of MSI

Name	Radius (MU)	Density (MU)	Surface temperature (K)	Escape velocity (MU)	MSI_I	MSI_S	MSI
Earth	1.88	1.40	288	2.23	0.70	0.66	0.68
Mars	1.00	1.00	240	1.00	1.00	1.00	1.00
Moon	0.51	0.85	197	0.48	0.77	0.66	0.71
Kepler-186 f	2.21	1.29	215	2.49	0.67	0.70	0.69
Kepler-438 b	2.11	1.27	312	2.36	0.72	0.60	0.66
Kepler-442 b	2.53	1.36	265	2.93	0.65	0.63	0.64

MU = Mars Units, where radius is 3390 km, density is 3.93 g/cm^3, escape velocity is 5.03 km/s.

The MSI is calculated and the sample is tabulated in Table 10.4, the calculation is computed using the weight exponent values of Table 10.3. Moreover, the scatter plot Figure 10.2, shows 12 Mars-like exoplanets (Kashyap et al. 2017).

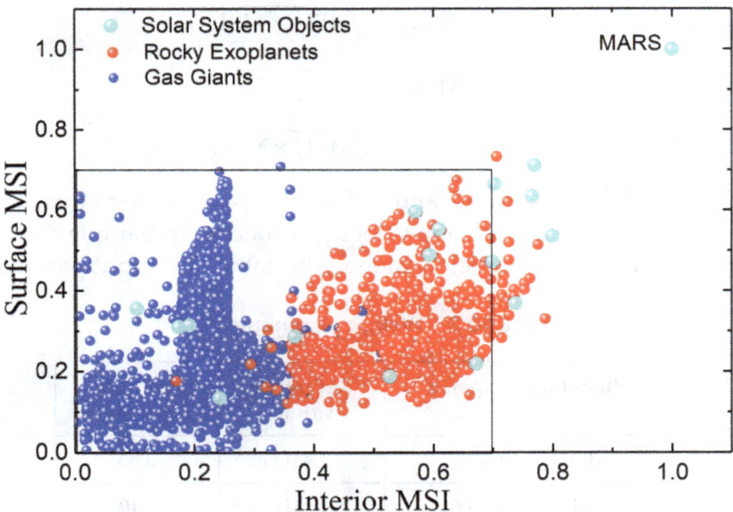

FIGURE 10.2 Interior MSI versus Surface MSI plot (for the updated figure refer Kashyap et al. 2017)

10.4 Tardigrades

The phylum Tardigrada (water bears or moss piglets) consists of over 1200 species (Guidetti and Bertolani 2005, Degma and Guidetti 2007, Degma et al. 2009) that inhabit aquatic (freshwater and marine) and terrestrial environments throughout the world, right from the deepest seas to the highest mountains (Ramazzotti and Maucci 1983, McInnes 1994, Nelson et al. 2010, 2015). Water bears are one of the toughest metazoans on Earth, and are often used in research on survivability in extreme conditions (Wright 2001, Guidetti et al. 2012). It has been shown that many tardigrade species have significant resistance to many physical and chemical environmental stressors like desiccation, very high and extremely low temperatures, high levels of ionizing radiation, high pressures and chemicals such as ethanol, carbon dioxide, hydrogen sulfide, 1-hexanol and methyl bromide gas (Baumann 1922, Becquerel 1950, Ramløv and Westh 1992, 2001, Seki and Toyoshima 1998, Jönsson and Guidetti 2001, Horikawa et al. 2006, Jönsson 2007, Ono et al. 2008, Wełnicz et al. 2011, Guidetti et al. 2012). Tardigrades are also able to survive in open space with combined exposure to space vacuum, and solar and cosmic radiation (Jönsson et al. 2008). Tardigrades owe this remarkable resistance to extreme conditions to their ability to enter into cryptobiosis. In cryptobiosis, metabolic processes significantly decrease (Pigoń and Węglarska 1955, Clegg 1973). Entering cryptobiosis and then returning to active life requires preparation that includes: a) forming a tun (Baumann 1922); and b) synthesizing many different molecules and bioprotectants such as non-reducing sugars (e.g. trehalose), late embryogenesis abundant proteins, heat shock proteins, cytoplasmic abundant heat-soluble proteins, secretory abundant heat-soluble proteins, mitochondrial abundant heat-soluble proteins and aquaporin proteins (Hengherr et al. 2008, 2009, Förster et al. 2009, Wełnicz et al. 2011, Guidetti et al. 2011, 2012, Yamaguchi et al. 2012, Grohme et al. 2013). Tardigrades also have very efficient DNA repair mechanisms (Rizzo et al. 2010, Wełnicz et al. 2011). In 2016, Hashimoto et al. discovered that tardigrade proteins could be used with human DNA to obtain protection from radiation. Tardigrade's active and cryptobiotic forms are used for understanding water extremophiles (see Figure 10.3).

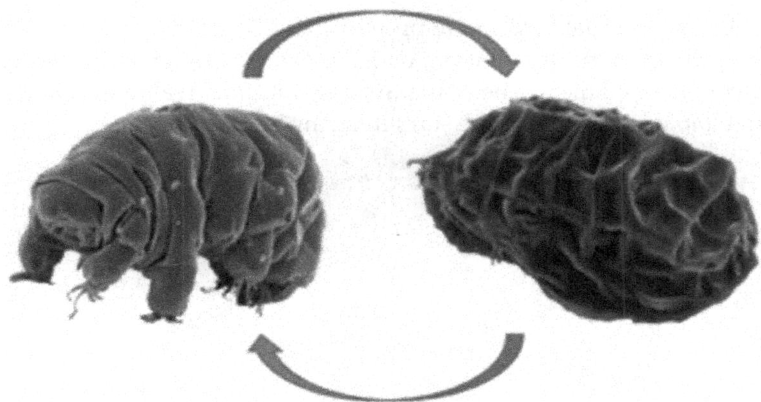

FIGURE 10.3 Tardigrades in active (left) and cryptobiotic (right) state (Credit: Beisser et al. 2018)

Since tardigrades obtain protection from radiation in space, it was important to theoretically check whether these extremophiles could survive in harsh conditions of exoplanets.

10.5 Active Tardigrade Index

ATI can be mathematically represented as (Kashyap et al. 2017):

$$ATI_X = [1-|(X - X_0)/(X + X_0)|]^{W_X} \tag{10.19}$$

where X is the physical parameter of the exoplanet such as radius (R), bulk density (ρ), escape velocity (V_e), planetary revolution (R_e), surface gravity, surface pressure or surface temperature (T_s), whereas X_0 is the same parameter with reference to Earth, and W_X is the weight exponent. These parameters are expressed as EU, while the surface temperature is represented in Kelvin (K).

The weight exponents for the upper and lower limits of the parameters were calculated following the method adopted by Schulze-Makuch et al. (2011): radius = 0.5 to 1.9 EU, mass = 0.1 to 10 EU, density = 0.7 to 1.5 EU, escape velocity = 0.4 to 1.4 EU, surface temperature and pressure (T = 1 to 38°C and P = up to 7500 Mpa for active, and T = –253 to 153°C and P = up to 7500 Mpa for cryptobiotic tardigrades). Similarly, we defined the limits of gravity as 0.16 to 17 EU, and planetary revolutions as 0.61 to 1.88 EU. Gravity and planetary revolutions are newly introduced weight exponents, which can be calculated from the following conditions: the human centrifuge experiment (Brent et al. 1960) clearly showed that untrained humans can tolerate 17 EU with eyeballs. In 2016, Hashimoto et al. discovered that tardigrade protein could be used on human DNA to obtain protection from radiation, which is a key component for tardigrade survival on exoplanets. The planetary revolution is scaled on the basis of the habitable zone of Sun-like stars in terms of Earth years.

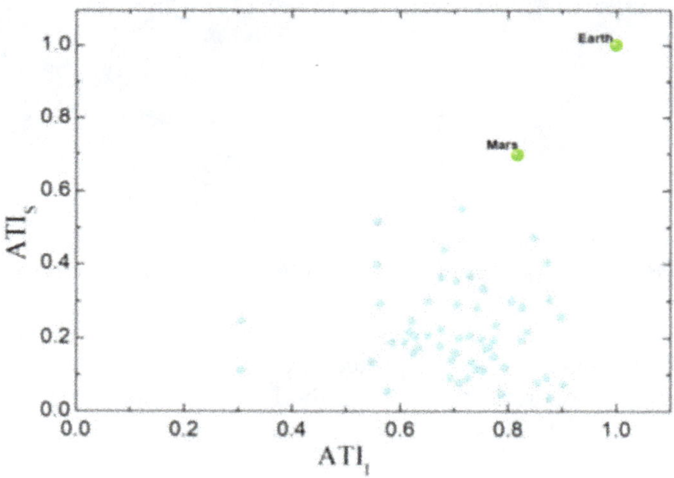

FIGURE 10.4 Scatter plot of Active Tardigrade Index interior vs surface

The results of ATI from Kashyap et al. (2018) found no planets close to Mars ATI value as evident from Figure 10.4. But the number may change with new discoveries of exoplanets in the future.

10.6 Cryptobiotic Tardigrade Index

The global Cryptobiotic Tardigrade Index can be mathematically represented as:

$$CTI_X = [1-|(X - X_0)/(X + X_0)|]^{W_X} \quad (10.20)$$

where X is the physical parameter of the exoplanet such as radius (R), bulk density (ρ), escape velocity (V_e), planetary revolution (R_e), surface gravity, surface pressure, and surface temperature (T_s), whereas X_0 is with reference to Earth, and W_X is the weight exponent (see Table 1). These parameters are expressed in the EU, while the surface temperature is represented in K.

The global CTI is divided into interior (CTI) and surface (CTI) which are expressed as:

$$\text{CTI} = \sqrt{\text{CTI}_I \times \text{CTI}_S} \quad (10.21)$$

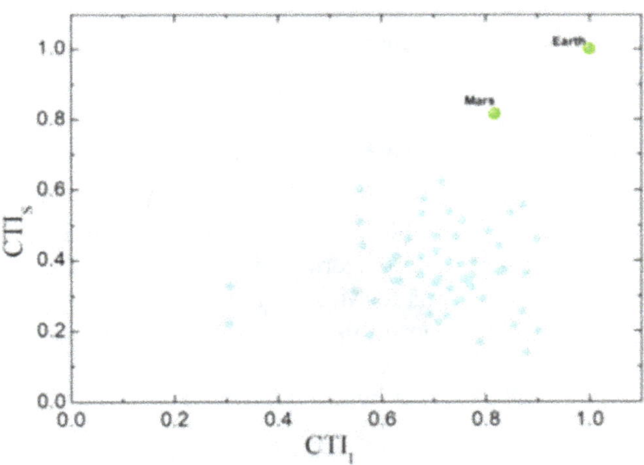

FIGURE 10.5 Scatter plot of Cryptobiotic Tardigrade Index interior vs surface

The results of CTI from Kashyap et al. (2018) found no planets close to Mars CTI value as is seen from Figure 10.5. But the number may change in view of new discoveries of exoplanets in the future.

The results of ATI and CTI from Kashyap et al. (2018) found no planets close to Mars ATI and CTI value as seen from Figure 10.6. But this number may change with new exoplanets being discovered in the future.

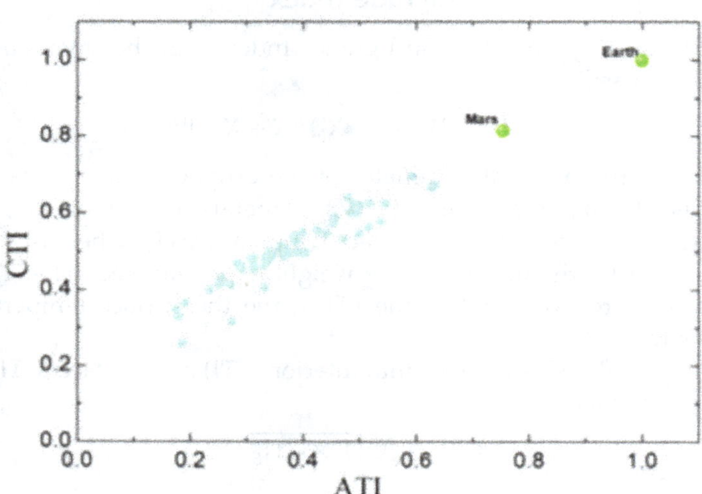

FIGURE 10.6 Scatter plot of Active Tardigrade Index vs Cryptobiotic Tardigrade index

10.7 Rock Similarity Index

Rock Similarity Index or RSI is designed to index Earth-like planets with physical conditions which, though harsh, are at least potentially suitable for rock-dependent extremophiles such as Chroococcidiopsis and Acarospora. According to Mckay (2014), the temperature range in which extremophilic microorganisms are able to reproduce and grow is between 258 K and 395 K. The corresponding weight exponent for surface temperature was calculated to be 2.26 (Kashyap et al. 2020).

$$RSI_X = [1 - |(X - X_0)/(X + X_0)|]^{W_X} \quad (10.22)$$

The weight exponents for the upper and lower limits appeared similar to the tardigrade indexes as per Kashyap et al. (2017), with the exception of surface temperature, which is adjusted to rocky extremophile. In order to calculate the weight exponent, the following upper and lower limits were used for each parameter: mean radius = 0.5-1.9 EU; bulk density = 0.7-1.5 EU; escape velocity = 0.4-1.4 EU; surface temperature T = 258-395 K; and revolution = 0.61-1.88 EU. The weight exponents were calculated applying these limits in the weight exponent equation previously established by Kashyap et al. (2018).

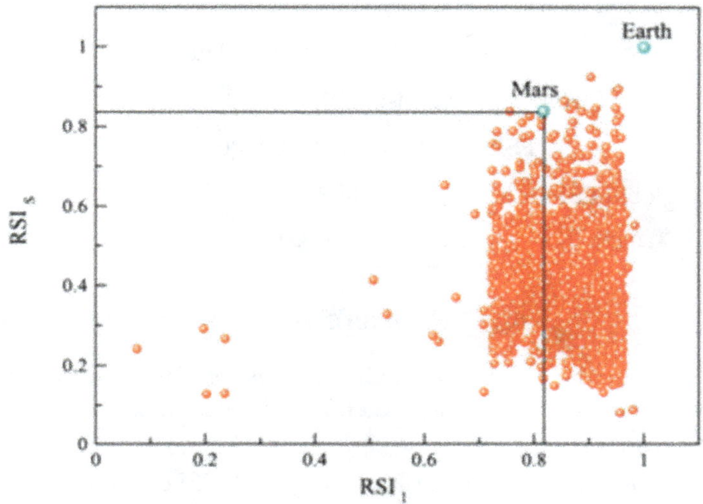

FIGURE 10.7 Scatter plot of Rocky Similarity Index interior vs surface (Kashyap et al. 2020)

As per the results of RSI shown in Figure 10.7, 21 confirmed potentially habitable exoplanets are seen (Kashyap et al. 2020). But this number keeps changing with the new discoveries of exoplanets every year.

Worked Example

To find the ESI of Mars

ESI calculations for Mars are performed using Eq. (10.10), with weight exponents from Table 10.1 using the following values for the input parameters,

$$R = 0.53 \times 6371 \text{ km} = 3376.63 \text{ km},$$
$$\rho = 0.71 \times 5.51 \text{ g/cm}^3 = 3.9121 \text{ g/cm}^3,$$
$$V_e = 0.45 \times 11.19 \text{ km/s} = 5.0355 \text{ km/s},$$
$$T_s = 240 \text{ K}$$

The ESI for each parameter is accordingly,

$$\text{ESI}\,R = (1 - |3376.63 \text{ km} - 6371 \text{ km}| / |3376.63 \text{ km} + 6371 \text{ km}|)\,0.57 = 0.8124,$$
$$\text{ESI}\rho = (1 - |3.9121 \text{ g/cm}^3 - 5.51 \text{ g/cm}^3| / |3.9121 \text{g/cm}^3 + 5.51 \text{ g/cm}^3|)\,1.07 = 0.8218,$$
$$\text{ESI}\,V_e = (1 - |5.0355 \text{ km/s} - 11.19 \text{ km/s}| / |5.0355 \text{ km/s} + 11.19 \text{ km/s}|)\,0.7 = 0.7162,$$
$$\text{ESI}\,T_s = (1 - |240 \text{ K} - 288 \text{ K}| / |240 \text{ K} + 288 \text{ K}|)\,5.58 = 0.587.$$

Interior ESI from Eq. (10.12) is:

$$\text{ESII} = \sqrt{0.8124 \times 0.8218} \approx 0.8171$$

Surface ESI from Eq. (10.13) is:

$$\text{ESIS} = \sqrt{0.7162 \times 0.5875} \approx 0.6487$$

The global ESI for Mars (Eq. 10.14) is:

$$\text{ESI} = \sqrt{0.8171 \times 0.6487} \approx 0.728$$

Exercise

1. Calculate the ESI radius of GJ 667Cc exoplanet, with radius weight exponent of 0.57 and the planetary radius of GJ 667Cc as 1.54 EU.

   Answer: 0.87

2. Calculate the interior, surface and global MSI of Proxima Centauri *b* exoplanet. The radius, density, escape velocity and surface temperature values are available in Table 10.3 in the MU. Compare and comment on the calculated global MSI with the global ESI value.

   Answer: MSI = 0.71 and ESI = 0.92

3. Calculate global ESI of a gas giant Jupiter and a rocky planet Mars. The required data is available in NASA planetary fact sheets. Conclude by commenting on the obtained results.

   Answer: ESI (Jupiter) = 0.21 and ESI (Mars) = 0.73

4. Calculate the weight exponents for ATI and CTI with threshold value 0.8.

   Hint: Use the upper and lower limits given in section 10.5 and 10.6

   Answer: Refer Kashyap et al. (2018)

5. Calculate the weight exponents of RSI with threshold value 0.8.

   Hint: Use the upper and lower limits given in section 10.7

   Answer: Refer Kashyap et al. (2020)

CHAPTER 11

Astrobiology

"Everyone is an astrobiologist; they just don't know it yet."

– Mary Voytek, NASA

11.1 Introduction

Astrobiology was formerly known as exobiology, and the word astrobiology was coined by the Russian astronomer Gavriil Tikhov in 1953 (Cockell et al. 2001). Astrobiology is an emerging multidisciplinary study of planetary formation, origin of life and involves search for extraterrestrial life forms. Astrobiology, in particular, is the study of life in the universe and serves as the baseline to study life in outer worlds beyond our Earth. Extremophile studies in space is the current line of research in astrobiology. "Extremophiles" are those microbes which love extreme environments such as very high or very low temperatures. Based on the observations from the evolution of earth, the limitations of survival characteristics are defined and compared with conditions prevalent on exoplanets (Hegde and Kaltenegger 2013).

11.2 Microorganisms Tested in Outer Space

Bacteria

Actinomyces erythreus (Simulated conditions, Dublin and Volz 1973), *Aeromonas proteolytica* (Low Earth orbit, Taylor et al. 1975), *Anabaena cylindrica* (akinetes) (Low Earth orbit and simulated conditions, Olsson-Francis et al. 2009), *Azotobacter chroococcum* (Simulated conditions, Moll and Vestal 1992), *Azotobacter vinelandii* (Simulated conditions, Roberts and Wynee 1962), *Bacillus cereus* (Simulated condition, Hagen et al. 1967), *Bacillus megaterium* (Simulated condition, Hawrylewicz et al. 1962), *Bacillus mycoides* (Simulated condition, Imshenetskiĭ et al. 1984), *Bacillus pumilus* (Simulated condition, Imshenetskiĭ 1984, Horneck 2012), *Bacillus subtilis* (Low Earth orbit, impact event and planetary ejection, atmospheric reentry, simulated conditions, Wassmann 2012), *Bacillus thuringiensis* (Low Earth orbit, Taylor et al. 1975), *Carnobacterium* (Simulated condition, Nicholson et al. 2012), *Chroococcidiopsis* (Low Earth orbit impact event and planetary ejection atmospheric reentry simulated conditions, Cockell et al. 2011), *Clostridium botulinum* (Simulated condition, Hawrylewicz et al. 1962), *Clostridium butyricum* (Simulated condition, Koike 1996), *Clostridium celatum* (Simulated condition, Koike 1996), *Clostridium mangenotii* (Simulated condition, Koike 1996), *Clostridium roseum* (Simulated condition, Koike 1996), *Deinococcus aerius* (Low Earth orbit, Kawaguchi et al. 2018), *Deinococcus aetherius* (Low Earth orbit, Yamagishi et al. 2018), *Deinococcus geothermalis* (Low Earth orbit and simulated condition, Neuberger 2015), *Deinococcus radiodurans* (Low Earth orbit, impact event and planetary ejection and simulated conditions, De La Vega et al. 2007), *Enterobacter aerogenes* (Simulated condition, Young et al. 1964), *Escherichia coli* (Low Earth orbit, impact event and planetary ejection and simulated conditions, Willis et al. 2006), *Gloeocapsa* (Low Earth orbit, Cockell et al. 2011), *Gloeocapsopsis pleurocapsoides* (Simulated condition, de Vera et al. 2013), *Haloarcula-G* (Low Earth orbit, Mancinelli et al. 1998), *Hydrogenomonas eutropha* (Low Earth orbit, Grigoryev et al. 1972), *Klebsiella pneumoniae* (Simulated condition, Hawrylewicz et al. 1962), *Kocuria rosea* (Simulated condition, Imshenetskiĭ et al. 1979), *Lactobacillus plantarum* (Simulated condition, Hawrylewicz et al. 1968), *Leptolyngbya* (Simulated condition, de Vera et al. 2013), *Luteococcus japonicus* (Simulated condition, Zhukova and Kondratyev 1965), *Micrococcus luteus* (Low Earth orbit, Zhukova and Kondratyev 1965), *Nostoc commune* (Low Earth orbit and simulated condition, Jänchena et al. 2015), *Nostoc microscopicum* (Simulated condition, de Vera et al. 2013), *Photobacterium* (Simulated condition, Zhukova and Kondratyev 1965), *Pseudomonas aeruginosa* (Simulated condition and low to Earth, Hawrylewicz et al. 1968), *Pseudomonas fluorescens* (Simulated condition, Hawrylewicz et al. 1968), *Rhodococcus erythropolis* (Low to Earth, Burchell 2001), *Rhodospirillum rubrum* (Simulated condition, Roberts and Wynee 1962), *Salmonella enterica* (Simulated condition, Raktim et al. 2016), *Serratia marcescens* (Simulated condition, Hagen et al. 1967), *Serratia plymuthica* (Impact event and planetary ejection, Roten et al. 1998), *Staphylococcus aureus* (Simulated condition, Hawrylewicz et al. 1968), *Streptococcus mutans* (Simulated condition, Koike et al. 1995), *Streptomyces albus* (Simulated condition, Hawrylewicz et al. 1968), *Streptomyces coelicolor* (Simulated condition, Koike et al. 1995), *Synechococcus (halite)* (Low to Earth, Mancinelli 2015), *Synechocystis* (Low to Earth and simulated condition, Klementiev et al. 2019), *Symploca* (Simulated condition, de Vera et al. 2013) and *Tolypothrix byssoidea* (Simulated condition, de Vera et al. 2013).

Archaea

Halobacterium noricense (Simulated conditions, Stan-Lotter 2002), *Halobacterium salinarum* (Simulated conditions, Koike et al. 1995), *Halococcus dombrowskii* (Simulated conditions, Stan-Lotter 2002), *Halorubrum chaoviatoris* (Low Earth orbit, Mancinelli 2015), *Methanosarcina sp.* (Simulated conditions, Morozova et al. 2006), *Methanobacterium* (Simulated conditions, Morozova et al. 2006), and *Methanosarcina barkeri* (Simulated conditions, Morozova et al. 2006).

Fungi and Algae

Aspergillus niger (Simulated condition, Zhukova and Kondratyev 1965), *Aspergillus oryzae* (Low Earth orbit and simulated condition, Dose 1995, Zhukova and Kondratyev 1965), *Aspergillus terreus* (Simulated condition, Sarantopoulou et al. 2011), *Aspergillus versicolor* (Low Earth orbit, Novikova et al. 2015), *Chaetomium globosum* (Low Earth orbit and simulated conditions, Taylor et al. 1975), *Cladosporium herbarum* (Simulated conditions, Sarantopoulou et al. 2014), *Cryomyces antarcticus* (Low Earth orbit and Simulated condition, Wall 2016, Claudia Pacelli et al. 2017), *Cryomyces minteri* (Low Earth orbit and Simulated conditions, Wall 2016), *Euglena gracilis* (Low Earth orbit and Simulated conditions, Häder et al. 2006, Nasir et al. 2014, Strauch et al. 2018, Strauch et al. 2010), *Mucor plumbeus* (Simulated condition, Zhukova and Kondratyev 1965), *Nannochloropsis oculata* (Impact event and planetary ejection, Pasini et al. 2015, 2013), *Penicillium roqueforti* (Low Earth orbit, Hotchin et al. 1965), *Rhodotorula mucilaginosa* (Simulated condition, Zhukova and Kondratyev 1965), *Sordaria fimicola* (Low Earth orbit, Zimmermann et al. 1994), *Trebouxia* (Simulated condition, Sánchez et al. 2013), *Trichoderma koningii* (Low to Earth 2013), *Trichoderma longibrachiatum* (Low Earth orbit, Neuberger et al. 2015), *Trichophyton terrestre* (Low Earth orbit, Taylor et al. 1975), *Ulocladium atrum* (Atmospheric reentry, Brandstätter 2008).

Lichens

Aspicilia fruticulosa (Low Earth orbit and simulated condition, Raggio 2011), *Buellia frigida* (Simulated condition, Meeßen et al. 2015), *Circinaria gyrosa* (Low Earth orbit and Simulated condition, Meeßen et al. 2015, Rosa et al. 2017), *Rhizocarpon geographicum* (Low Earth orbit and Simulated condition, Meeßen et al. 2015, de La Torre Noetzel 2007), *Rosenvingiella* (Low Earth orbit, Cockell et al. 2011), *Xanthoria elegans* (Low Earth orbit, impact event and planetary ejection and simulated condition, Sancho 2007, De Vera et al. 2004, Horneck 2008, Brandt et al. 2014, Horneck 2008), and *Xanthoria parietina* (Low Earth orbit and simulated condition, Brandt et al. 2014).

Bacteriophage/Virus

T7 phage (Low Earth orbit and Simulated condition, Taylor et al. 1975), *Canine hepatitis* (Low Earth orbit, Hotchin 1968), *Influenza PR8* (Low Earth orbit, Hotchin 1968), *Tobacco mosaic virus* (Low Earth orbit, Hotchin 1968), and *Vaccinia virus* (Low Earth orbit, Hotchin 1968).

Yeast

Rhodotorula rubra (Low Earth orbit and simulated condition, Taylor et al. 1975), *Saccharomyces cerevisiae* (Low Earth orbit and simulated condition, Taylor et al. 1975), *Saccharomyces ellipsoides* (Low Earth orbit, Grigoryev 1972), and *Zygosaccharomyces bailii* (Low Earth orbit, Grigoryev 1972).

Animals

Caenorhabditis elegans [nematode] (Low Earth orbit, Higashibata 2006), *Hypsibius dujardini* (tardigrade) (Low Earth orbit, impact event and planetary ejection, Pasini et al. 2014), *Milnesium tardigradum* (tardigrade) (Jönsson et al. 2008), *Richtersius coronifer* (tardigrade) (Low Earth orbit and simulated condition, Jönsson and Wojcik 2017), and *Mniobia russeola* (Simulated condition, Jönsson and Wojcik 2017).

11.3 Habitability Models in Astrobiology

Habitability is a nomenclature commonly found in planetary science and astrobiology literature, indicating if an environment can or cannot sustain life of a particular form (Cockell et al. 2016). A decade back, planetary scientists proposed many proxy habitability models or indices for the Earth, Mars, the solar system and other extrasolar bodies (e.g. Stoker et al. 2010, Schulze-Makuch et al. 2011, Armstrong et al. 2014, Barnes et al. 2015, Silva et al. 2017, Kashyap Jagadeesh et al. 2017, Rodríguez-López et al. 2019, Seales and Lenardic 2020) to understand habitability of planets. Also, there are some specific, universal biological quantities that can be used as proxies for habitability, such as carrying capacity, growth rate, metabolic rate, productivity, presence of certain requirements of life, or even genetic diversity (Heller 2020). There is also an ongoing debate as to whether any concept of habitability needs to be binary (yes/no) in nature as proposed by Cockell et al. (2019), or continuous as per Heller (2020), or probabilistic

(Catling et al. 2018). While a binary interpretation of habitability only allows a given planet to be habitable to a given species, or not, a continuous model allows for the possibility of a world (planet or moon) to be more habitable than Earth, that is, to be superhabitable (Heller and Armstrong 2014). A direct measure of habitability requires the knowledge of how the environment affects at least one of the biological quantities pertaining to a given species or community. We do not need to specifically estimate these quantities, but only know how the environment proportionally affects them. For example, we know how temperature affects the productivity of primary producers such as plants and phytoplankton. Most require temperatures between 0° and 50°C, but such producers do better (i.e. have the highest productivity) close to 25°C (Silva et al. 2017). Their 'thermal habitability function' looks like a Gaussian distribution curve centered about their optimum productivity temperature. Habitability measured directly is better represented as a fraction from zero to one.

Biological productivity refers to the dry or carbon biomass produced over space and time. It is one of the best habitability proxies, which can be easily estimated for many ecosystems via ground or satellite observations. The first global-scale empirical model is the *Miami Model,* which gives a good approximate estimation of terrestrial Net Primary Productivity (NPP), the rate of photosynthetic carbon fixed minus the carbon used by autotrophic respiration (Zaks et al. 2007). This simple model only uses two factors, annual mean surface temperature and precipitation, to successfully infer global distribution of vegetation (Adams et al. 2004). One important limitation of this type of model is that climate variables such as precipitation not only affect crops but are also in turn affected by vegetation; e.g. there is increasing evidence that tropical forests have a strong impact on precipitation patterns on Earth (Molina et al. 2019). Nowadays, different complex biogeochemical models and satellite observations (e.g. NASA's TERRA, AQUA, and Soumi NPP models) are combined to calculate local to global NPPs (Cramer et al. 1999, Ito et al. 2011). Satellite data is used to arrive at habitability indices to monitor the current terrestrial biodiversity and the data is also used through climate change (e.g. Pan et al. 2010, Radeloff et al. 2019). Thus, NPP is also a measure of global terrestrial health or habitability since primary producers form the bases of the food chain.

Most habitability models are limited to indirect measures of habitability due to lack of information. This is especially true for extrasolar planets (exoplanets). For example, the availability of Earth-size exoplanets in the Habitable Zone of stars (termed the *Eta-Earth* value) can be considered an indirect measure of stellar habitability (i.e. the suitability of stars for habitable planets around them). Habitable Zone, the region around a star where an Earth-like planet could maintain surface liquid water, is generally considered to be a binary indirect measure of planetary habitability (Kasting et al. 1993), although many have argued it should be considered a probability density function (Zsom 2015, Catling et al. 2018). Though the location of the Habitable Zone depends mainly on the stellar type, its extension depends on the physical properties of the planet, in particular on the planet's atmospheric response to the stellar energy flux it receives (Kane 2013). Hence, the presence of liquid water on the surface also depends on the planet's atmospheric dynamics, which effectively homogenize differential heating of the surface, creating a short-term response to the planet's global temperature. This differential heating is a result of the planet's obliquity, which governs the latitudinal distribution of incoming stellar radiation (Nowajewski et al. 2018).

The Habitable Zone can be defined in terms of either the planet's distance from the star, its incoming stellar flux or its global equilibrium temperature. With regard to equilibrium temperature, however, the extension of the Habitable Zone depends on the planet's orbital parameters, particularly its eccentricity and obliquity. For example, when the orbital eccentricity increases the average equilibrium temperature decreases, thus extending the size of the Habitable Zone (Méndez and Rivera-Valentín 2017). Similarly, higher fixed obliquity and/or rapid changes in obliquity result in higher average equilibrium temperatures, which also result in extending the outer edge of the Habitable Zone (Armstrong et al. 2014). Further, for equilibrium temperature consideration, the extension of the Habitable Zone depends ultimately on the planet's energy balance. On Earth, the global energy balance is a result of the complex interaction between physical and biological processes. Biota affects the global energy balance in mutliple ways, including direct effects on surface albedo and latent heat fluxes (e.g. transpiration) (Jasechko et al. 2013).

The Earth Similarity Index (ESI), inspired by the diversity-similarity indices used in ecology to compare populations (Boyle et al. 1990), is a measure of Earth-likeness for a selected set of planetary parameters (Schulze-Makuch et al. 2011). Future observational constraints of Earth-similar atmospheric constituents (i.e. N_2, CO_2, H_2O) could improve our understanding of this and similar metrics. For instance, 3D global climate models indicate that spectral features of water vapor in close-in terrestrial exoplanetary atmospheres may be detectable by the *James Webb Space Telescope* (Chen et al. 2019), depending on the presence of clouds (Komacek et al. 2020). Even though the presence of water vapor in the atmospheres of terrestrial exoplanets can indicate habitability, it is necessary to perform detailed and exhaustive work to determine the species that can survive under conditions of extreme humidity. For example, mammals are not capable of surviving hyperthermia resulting from high air temperatures and highly humid conditions. So, planets with extreme differential heating across latitudes may be uninhabitable despite having liquid water on their surface (Nowajewski et al. 2018).

The Habitable Zone was developed to find terrestrial exoplanet targets that could potentially host life. It was first proposed by Edward Maunder in 1913 (Maunder 1913, Lorenz 2020) in his book *Life on Other Planets* with refined definitions later on (Huang 1959, Hart 1978, Kasting et al. 1993, Underwood et al. 2003, Selsis et al. 2007, Kaltenegger and Sasselov 2011, Kopparapu et al. 2013, 2014, Ramirez and Kaltenegger 2017, 2018). The general definition of a Habitable Zone currently in use is:*It is the circumstellar region around a star where a terrestrial planet with a suitable atmosphere could host liquid water on its surface.* The insistence on the presence of liquid water on the surface is essential since life on Earth requires liquid water to sustain. This definition is adequate only for remote observation of planets without considering any life that might exist at the subsurface. The reason being: the search for life on exoplanets will rely on remote observations of atmospheres, lacking the sophistication of *in-situ* measurements used in solar system planetary science. Therefore, identifying water in the atmosphere of exoplanets (in addition to other biosignature relevant gases) is the only way to narrow down potential life-hosting targets, as subsurface life deep in the interior may not be able to modify the atmospheres of planets enough to be remotely detectable. The current Habitable Zone paradigm has a specific application and is not inappropriate in neglecting the subsurface oceans in the outer solar system, the Venus clouds, or other environments far from Earth-like conditions. The Habitable Zone does not tell us if there are planets in that zone and if planets are present, whether they are habitable or not, but it only shows the impact of a few important variables on planetary habitability.

Abundance of liquid water in a planetary environment may be inherently unstable (Gorshkov et al. 2004), which leads to questions about the role of life in the definition of habitability itself (Zuluaga et al. 2014). Thermodynamic in-equilibrium may be one the most conspicuous signatures of a habitable (and inhabited) planet (Kleidon 2012). One common problem with some, if not all biological models, is they assume that climate (more generally, the full gamut of physical characteristics of the environment) is *a boundary condition* for life, implying that biological systems depend on climate and not the other way around. This premise is challenged by the fact that the observed state of the Earth system is the result of a complex and dynamic interaction between biological (e.g. ecosystems) and physical (e.g. climate) systems (Budyko 1974, Gorshkov et al. 2000, Kleidon 2012, Zuluaga et al. 2014). A critical question is how such a state, which is thermodynamically unstable (e.g. the composition of Earth's atmosphere) (Lenton 1998, Kleidon 2012), can be maintained over eons despite variables like solar luminosity and sudden large external thrusts like asteroid impacts? The answer depends on the interactions between biological and physical systems on Earth. A planet might be habitable, its state becoming compatible with the presence of liquid water during a given period of time just *by chance* (a random change in the planetary energy balance due to any combination of variables), but long-term persistence of its habitable state indicates the existence of natural regulation mechanisms (Walker et al. 1981, Lenton 1998, Gorshkov et al. 2000, Salazar and Poveda 2009), e.g. how Earth has maintained its habitable state for around 4 billion years.

Habitability model can be constructed in six steps: (1) Select the space and time of the region of interest (i.e. define the boundary conditions); (2) select variables and convert them to quantities proportional to mass or energy; (3) select species or communities and their ecophysiological response curves for the selected variables; (4) identify one or more standards of comparison (i.e. a terrestrial or planetary analog); and (5) solve the habitability master equation (Méndez et al. in preparation):

$$\frac{\partial^2 H}{\partial s \partial t} = \left(\frac{1}{M} \frac{\partial^2 M}{\partial s \partial t} + \frac{1}{E} \frac{d^2 E}{\partial s \partial t} \right) H + \frac{\partial M}{\partial s} \frac{\partial E}{\partial t} + \frac{\partial M}{\partial t} \frac{\partial E}{\partial s} \tag{11.1}$$

where H is the habitability, M and E are the normalized mass and energy, respectively, available (i.e. fraction of the total mass and energy) for the life of interest (i.e. a species or community) relative to the standard of comparison, s and t are the spatial (e.g. area or volume) and temporal (e.g. hours or days) components. The last step is: (6) validate and correct the habitability model with environments on Earth if possible (i.e. find positive correlations between habitability and biomass, productivity, or biosignatures). For example, equation 11.1 can be used to compare the habitability of a specific volume of ocean water of Europa relative to the same volume of deep ocean waters on Earth, knowing mass and energy fluxes. Sometimes it is not desirable to use the same volume or time periods for comparison purposes (e.g. comparing early Mars with contemporary Earth). The space of interest is not limited by planetary scales. It can be enclosing a stellar region to evaluate its overall habitability, e.g. a galactic habitable zone (Spitoni et al. 2017). The general population growth equations (e.g. exponential and logistic) can also be derived from equation 11.1.

The construction of a habitability model is not easy. The current model provides at least an upper limit for the habitability of a system for a given set of parameters, and is further improved by properly selecting key environmental variables and ecophysiological response curves for the life of interest (steps 2 and 3). Any model must be validated with environments on Earth, where a positive correlation between habitability and the presence or abundance of life is observed (step 6). The simple solution of equation 11.1 is $H = ME$ (units of kgJ), or more practically, $H' = \rho P$ (units of $Wkgm^{-3}$), where H' is the specific habitability, ρ is the concentration of one or more ingredients necessary for life, and P is the available metabolic power (assuming that mass depends only on space, and energy only on time). In practice, each of these variables could be normalized to the standard of comparison for simplicity and consistency with zero denoting a non-habitable environment and one denoting a highly habitable environment. Also, occurrence or probabilities could be used instead of these variables, which is exactly what the definition of the Habitable Zone is all about (i.e. probability of surface water ≥ 0). Negative values could be used to quantify the damaging effect of non-habitable environments (e.g. comparing the surfaces of the Moon and Venus). Values larger than one could represent super-habitable conditions relative to a standard.

As an example, general mass-energy habitability could be applied to calculate and compare the habitability of ocean worlds (e.g. Europa and Enceladus), the lakes of Titan, and the clouds of Venus. This is not an Earth-centric model since mass and energy are conserved quantities and no life could take more mass and energy than available in the environment. The general modeling approach is to start initially with simple models (e.g. fewer variables under steady-conditions) and then progress with complex dynamical models. A library is then created depending on the variables, the scales (e.g. site, regional, global) or life forms of interest. Likewise, there are several interesting terrestrial analogs that would be intriguing to look into with the proposed habitability models. Our models might be particularly well suited to look at low diversity and low biomass extreme environments. Also of interest are extreme environments supposedly sterile (e.g. the acidic brines of Dallol-Danakil or brines with a very high content of Mg or other chaotropes) (Belilla et al. 2019).

11.4 Scientific Queries

According to Méndez et al. (2021), future astrobiological missions should address basic scientific questions about the environment (s) to be studied. To do this, it is important to define an environment of interest, both in space and time.

1. **What are the limiting factors?** For confinement rules, there exist a small set of main factors (e.g. edaphic factors) which influence living organisms (e.g. water, nutrients). These are the first set of variables that have to be used for the construction of a habitability model, which will later be modified with more variables. For example, the primary variable is mainly driven by temperature, precipitation and nutrients on land, and by temperature and nutrient concentrations in the oceans, among other factors. In general, these variables should be directly or indirectly related to the mass and energy of the environment (Martiny et al. 2006, Pikuta et al. 2007, Williams and Hallsworth 2009, Harrison et al. 2013, McKay 2014, Lynch and Neufeld 2015, Tecon and Or 2017).
2. **What are the terrestrial and planetary analogs?** As a minimum requirement, identify at least one analog on Earth and another similar planetary analog as the comparison standard (i.e. for model normalizations). For example, if we are studying a particular Martian environment, choose the terrestrial polar deserts and a Martian analog based on the variables of interest. The cross-comparison of similar types of environment (e.g. salterns), as well as slightly different environments (e.g. high salinity biotopes with different pH, temperature or chemical conditions), can prove useful. The subsurface oceans of Europa or Enceladus can be compared with deep seawater, hydrothermal systems, or deep-sea brines (Méndez et al. 2021). Also, planetary atmospheres at high altitudes or near-space regions can be compared. Analysis of similarities (ANOSIM) could be used to formally choose and compare these regions (Clarke 1993).
3. **What is habitability value?** The habitability zone is evaluated on the chosen environmental factors and then compared with the selected Earth and planetary analogs (Example: Mars), using a normalized scale from zero to one for simplicity (Example: ESI and MSI). A collection of habitability measures is constructed (i.e. a habitability matrix), each for a different consideration (e.g. species). These inputs are then used to create multivariate habitability maps (*niche quantification* in ecology) for site selections.

4. **What is potential biomass?** The upper limits of biomass can be projected based on the fluxes of mass and energy available for life, which usually is a very small fraction of the total mass and energy. For example, biomass could be estimated from the available metabolic energy using the Metabolic Theory of Ecology (van der Meer 2006, Schramski et al. 2015, Clarke 2017). These upper limits are used in the designs of life detection experiments. Available free energy from the known disequilibria is used to estimate an upper limit on the biomass in the subsurface of Mars, and its value depends on uncertainties of the abundances of metabolic reactants and the assumed microbial basal power requirement (Sholes et al. 2019).
5. **What will be the expected correlation between habitability and biosignatures?** The potential upper values of biomass can be converted to estimate the observable biosignatures, also known as disequilibrium chemistry (Catling et al. 2018). Habitability and biosignatures are positively correlated on Earth but this need not be necessarily true for other planets. A zero or negative correlation could indicate an incorrect habitability model or a biological process different from that on Earth (*life as we don't know it*). The habitability-biosignatures correlation is a fundamental problem of astrobiology, but non-detections are also important. For example, it is very important to robustly detect planetary regions identified as habitable by Earth standards and yet devoid of any life. Such discoveries would place bounds on abiogenesis (i.e. the origin of life).

11.5 Extraterrestrial Intelligence

The term Extraterrestrial life is different from Extraterrestrial intelligence (ETI). The prospects of intelligent life in the universe is one of the ultimate questions of human curiosity. ETI can be defined as the living beings who are superior and advanced than human beings with respect to technology. This question is purely based on probability work performed by various bright minds for centuries. The Copernican revolution started the insights, when Copernicus discovered the Earth was a planet revolving around the Sun, and other planets were conversely, other worlds, which generalized the relativistic concept that the humans are not privileged observers of the universe (Peacock 1998). In the modern era, Stephan Hawking proposed the sheer scale of the universe, which makes it improbable for intelligent life not to have emerged elsewhere (Hickman 2010). Moreover, Fermi's Paradox gives the contradiction between high estimates of the probability of the existence of extraterrestrial civilization and humanity's lack of contact with, or evidence for, such civilizations (Krauthammer 2011). Kardashev Scale is one more way of measuring a civilization's level of technological advancement, based on the amount of energy a civilization is able to utilize (Kardashev 1964).

11.6 Kardashev Scale

In (Kardashev 1964), Russian astrophysicist Kardashev proposed a hypothetical method of understanding how technically advanced a civilization is depending on the amount of energy it can harness. Accordingly, he defined three types of civilizations based on the order of magnitude of power available to each of them.

Type I: Type I civilization, also known as Planetary civilization, is characterizedbythefactthattheplanetcanstoreandharnesstheentireenergy received by it from the host star. 'For example, the Sun' emits approximately 4×10^{26} watt of energy into space or 4×10^{26} joule of energy every second. But the Earth receives around 1360 watt per square meter of area, equivalent to 47% of the Sun's power. Even 5% of this energy is more than 50 times our energy requirement. Technological level of a Type I civilization is close to the current level of 4×10^{12} watt achieved by the Earth. American theoretical physicist Michio Kaku, however, feels that it will take a minimum of 100-200 years more for the Earth to reach the status of planetary civilization. As per Carl Sagan, we are 70% close to Type I. A Type I civilization not only colonizes the entire surface of the planet but also Low Earth Orbits (LEO).

Type II: Type II civilization, also known as stellar civilization, is capable of harnessing and controlling the entire energy radiated by its host star. In 1960, British-American theoretical and mathematical physicists put forth the idea of a hypothetical megastructure, now famous as the Dyson sphere, around a star and encompassing it. Type II civilization is capable of building such a sphere that would capture the entire energy emitted by the star and make it available for use by the home planet. Lemarchand defined this as a civilization capable of applying and mobilizing the entire radiation of the star's output. If the Earth was inhabited by stellar civilization, the energy use would then be matching the luminosity of the Sun, i.e. 4×10^{26} watt. The Earth would require another 1000-2000 years to reach this level of civilization.

Type III: This is a civilization which can rein in and control energy of the magnitude as the entire galaxy in which it is located and is hence aptly known as galactic civilization. Energy consumption of this civilization will be around $\approx 4 \times 10^{37}$ watt. According to Lemarchand, this type of civilization consumes power comparable to the luminosity of our entire Milky Way galaxy. It would probably take 100,000 years for us to reach this level and by then human life would no longer be present on the Earth. Though difficult to characterize, many theorists point out that a Type III civilization would be capable of building megastructures similar to the Dyson sphere, encompassing the core of the parent galaxy or the entire galaxy!!!

Kardashev did not go beyond Type III. American astronomer Carl Sagan interpolated and extrapolated the Kardashev scale and defined the 1970 civilization on the Earth with an energy consumption rate of 10 TW as Type 0 civilization. Other astronomers have extended the Kardashev scale to Type IV civilization, also known as a universal civilization, that has the ability to consume energy at the same scale as its host universe, and Type V or Type Ω civilization with the capability to create universes with its energy consumption at the scale of multiple universes. Additional types of civilization are also on the anvil but beyond human comprehension.

11.7 Drake Equation

Mankind from time to time keeps looking at the unfathomable night sky and realizes that the Earth is miniscule and insignificant in the vastness of the universe. While still struggling to understand the universe at large, people have also wondered if they are alone in this universe! Assuming long distance communication is possible with extraterrestrial civilizations, is it possible to get in touch with each of them? Do we even know how many alien societies exist? In 1961, at a meeting of experts on the subject held in West Virginia, American astrophysicist Dr. Frank Drake put forth an intriguing equation to arrive at the probable number of intelligent civilizations N from whom the electromagnetic signals are detectable in our own Milky Way galaxy. The equation summarizes the factors on which our likelihood of detecting signals from intelligent extraterrestrial life depends.

The famous Drake equation is:

$$N = R \cdot f_p \cdot n_e \cdot f_l \cdot f_i \cdot f_c \cdot L$$

Here, R = rate of formation of stars (per year) suitable for the development of intelligent life

f_p = fraction of the above stars that have planetary systems
n_e = number of planets around each of these stars within an "ecoshell", an environment conducive for life
f_l = fraction of life sustaining planets on which life actually sustains
f_i = fraction of life supporting planets on which intelligent life or civilizations emerge
f_c = fraction of intelligent civilizations that have survived long enough to develop communication technology to emit detectable signals of their existence into space
L = the average time in years over which such civilizations emit detectable signals before they cease to exist

As per the original estimation of N using Drake equation (Drake and Sobel 1992), assuming,

(i) 1 star forming over one year in Milky Way galaxy, $R = 1\ yr^{-1}$.
(ii) One fifth to one half of all stars formed having planetary systems around them,

$$f_p = 0.2 \text{ to } 0.5$$

(iii) One to five planets per host star capable of supporting life, $n_e = 1$ to 5.
(iv) Life will develop on all of these planets, $f_l = 1$.
(v) Again intelligent life will develop on 100% of these planets, $f_i = 1$.
(vi) 10 to 20% of the civilizations will be able to communicate, $f_c = 0.1$ to 0.2.
(vii) Communicable signals last between 1000 to 100,000,000 planets years, $L = 1000$ to 100,000,000 years.

N turns out to be 20 when lower limits of the factors are used and 50,000,000 if upper limits are substituted in the equation. Given the uncertainties in each of the factors, Drake equation suggested N is approximately equal to L, implying there are probably 1000 to 100,000,000 in the Milky Way galaxy with civilizations.

Current estimates of N are based on latest calculations from NASA and ESA, recent analysis of microlensing surveys, data from Kepler Space Mission and suggestions by scholars on the subject. L, in particular, turns out to be more bullish, considering successive civilizations have evolved on Earth, for example, each communicating in a different mode, not necessarily technical.

All the hypotheses and uncertainties in the variables involved in Drake may project widely varying values for N. Of particular interest is the case where $N \lll 1$, indicative of we earthlings being alone in our galaxy. $N > 1$ implies that at least one extraterrestrial civilization exists and can be contacted!!!

11.8 Fermi Paradox

Put forth by the eminent Italian-American physicist Enrico Fermi, Fermi paradox is the discrepancy between the lack of evidence of extraterrestrial intelligent life and our estimated probability of existence of alien life. We see no signs of alien technology, and our radio technology doesn't pick up voices from other worlds (Adler 2020), though theoretically space is teeming with habitable planets. Fermi documented this paradox in 1950 but unfortunately passed away in 1954.

Michael Hart, an American astrophysicist, examined the Fermi paradox in detail and published the results in the quarterly journal of Royal Astronomical Society 1975 in a paper titled "An Explanation for the Absence of Extraterrestrials on Earth". This has become the reference point for researchers involved with the Fermi paradox, now more aptly known as Fermi-Hart paradox. Though Hart wrote in the abstract: "It is suggested that this fact can best be explained by the hypothesis that there are no other advanced civilizations in our galaxy", he did not fail to mention that more detailed research in the fields of biochemistry, planetary formation and atmospheres was needed to better narrow down the answer.

Several hypotheses have been proposed to resolve the Fermi-Hart paradox but with little success.

11.9 The Great Filter

It remains an unsolved mystery why no aliens have ever visited the Earth nor have we been able to detect signals from them, despite a conservatively miniscule but finite probability of their existence. Many theories have been put forth to account for the 'Great Silence' around us. Robin Hanson, a research associate at Oxford University's Future of Humanity Institute, coined the term "The Great Filter" to explain why the world around is dead! Hanson proposed that The Great Filter prevents abiogenesis – non-living matter getting together to form living organisms. Hanson also reasoned out that The Great Filter accounts for why civilizations cannot achieve high levels of development on the Kardashev scale.

Let us consider the various steps that are essential for a civilization to emerge as a space-faring civilization, colonizing most of the visible world.

1. A planet capable of supporting life should form in the habitable zone of the parent star.
2. Life must develop on the planet.
3. Developed life forms should be capable of reproduction using DNA and RNA molecules.
4. Simple cells (prokaryotes) must evolve into more complex cells (eukaryotes).
5. The resulting multicellular organisms must develop.
6. Genetic diversity should take place through sexual reproduction.
7. Complex organisms with big sized brains capable of handling tools should evolve.
8. The above organisms should develop advanced technology needed for industrial revolution and space colonization. Apparently, we earthlings are currently at this stage.
9. Colonization explosion should take place wherein the space-faring species should colonize other worlds and star systems without destroying themselves.

The Great Filter hypothesis makes one of the above steps improbable. As per Hanson, The Great Filter should be located somewhere between the origin of life on a planet and the point where the civilization becomes interplanetary or interstellar as shown in Figure 11.1 below:

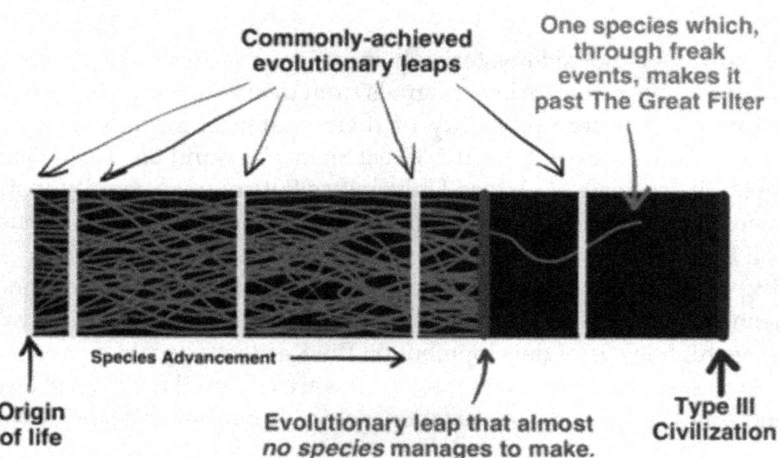

FIGURE 11.1 Schematic representation of The Great Filter hypothesis (Image credit: Tim Urban 2014)

There are two possibilities:

(i) The Great Filter is behind us, meaning we have passed it. This suggests it must have been difficult for life to emerge from inorganic matter in the early stages. The good news is we are the only species to have freaked out past The Great Filter. Not only Earth but the entire universe is ours for the taking. This loneliness we feel in this vast void is actually a blessing!

(ii) If The Great Filter is ahead of us, we will most likely detect at least one alien civilization and, more importantly, we are nearing the doomsday of our extinction. Perhaps our progress from step 8 to step 9 is, alas, not probable!

11.10 Interesting Discoveries

Life on Venus

As we know, Search for ExtraTerrestrial Intelligence (SETI) refers to a collective effort by scientific researchers to explore intelligent extraterrestrial life. SETI involves monitoring electromagnetic radiation mostly in the radio and visible range for possible signal transmission from civilizations outside the solar system. It is different from search for extraterrestrial life, which preempts there is no dearth of building blocks of life in the universe, and the laws of physics, chemistry and biology are valid throughout the cosmos. It is imperative that researchers look for signature conditions prevailing in solar system planets and exoplanets that are actually defined by the variety of terrestrial life forms (O'Callaghan 2020). Phosphine gas (PH3) consisting of molecules of one phosphorus and three hydrogen atoms is a biomarker since on Earth it is produced by microbes. In 2017, a team of researchers from Cardiff University, UK detected the presence of phosphine gas in the atmosphere of Venus using the James Clerk Maxwell Telescope, suggesting the presence of extraterrestrial life (Greaves et al. 2020). We quote from the research paper: "The presence of PH3 is unexplained after exhaustive study of steady-state chemistry and photochemical pathways, with no currently known abiotic production routes in Venus's atmosphere, clouds, surface and subsurface, or from lightning, volcanic or meteoritic delivery. PH3 could originate from unknown photochemistry or geochemistry, or, by analogy with biological production of PH3 on Earth, from the presence of life" (Greaves et al. 2020).

TYC 7037-89-1 Sextuple Star System

TYC 7037-89-1, also known as TIC 168789840, is a six-star system of 3 groups of eclipsing binaries in the constellation of Eridanus, which was discovered using NASA's TESS telescope and reported by Powell et al. on 9th January, 2021. This star system located close to 2000 light years away is the first one to consist of three stellar binaries, all orbiting around a common barycenter. The component stars cannot be generally resolved using a telescope. Scientists were lucky to see the spot of light due to this sextuple system since all the stars orbited in a plane in sync with their line of observation. Observers noted the intriguing dimming and brightening of the spot of light as the stars kept eclipsing each other.

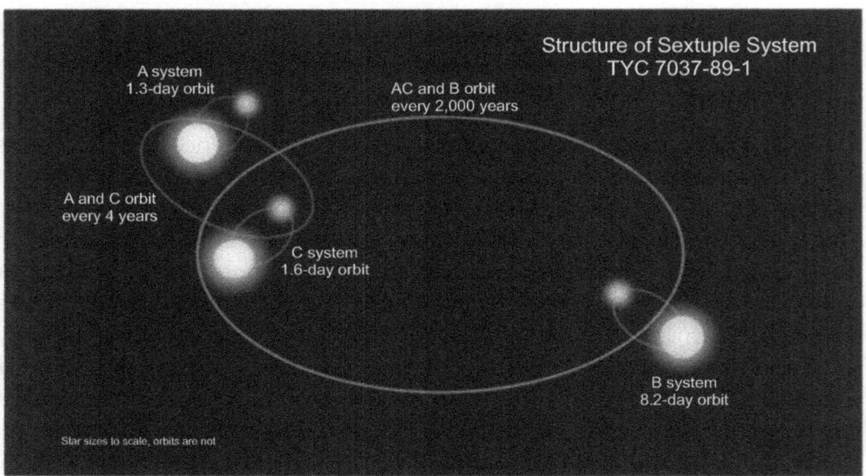

FIGURE 11.2 Schematic representation of the sextuple star system TYC 7037-89-1 (Credit: Jeanette KazmierczakNASA's Goddard Space Flight Center)

The astronomers have designated the binaries as A, B and C (refer Figure 11.2). The stars in these binaries orbit each other in 1.3, 8.2 and 1.6 days, respectively. A and C binaries orbit each other every four years. The primary stars in all the three binaries are all slightly bigger and more massive than the Sun and about as hot. The secondary companions are all around half the Sun's size and have a third of Sun's effective temperature (Powell et al. 2021).

TYC 7037-89-1 joins the league of Castor, another sextuple system in the constellation of Gemini, along with 16 other similar systems as per June, 2020 update.

Septuplets Around TRAPPIST-1

The red dwarf star TRAPPIST-1, located around 40 light-years away, was first discovered in 1999. In 2016, Transiting Planets and Planetesimals Small Telescope (TRAPPIST) spotted three planets in the habitable zone of this star. In honor of this telescope, the star came to be called TRAPPIST-1. By 2017, NASA's Hubble telescope and now retired Spitzer Space Telescope discovered four more planets around TRAPPIST-1 to join the clan. These septuplets are so close to the host planet that the entire planetary system, twice as old as our solar system, could be accommodated within the orbit of Mercury! The farthest planet has an orbital period of just 19 days! If you were to stand on any one of these fascinating planets, you would perhaps see the other planets hovering above! Continuous observation of the star light from TRAPPIST-1 enabled astronomers to precisely arrive at the septuplets' mass, diameter and hence densities in 2021. It turns out that all of them have comparable density not varying by more than 3%. This is a pointer to their similar rocky composition and hydrogen dominated atmosphere.

TrES-2b, the Darkest Alien Planet

TrES-2b, a planet located some 750 light years away in the constellation of Draco orbiting around a yellow main sequence star GSC 03549-02811, is the darkest exoplanet on record and is barely visible. It reflects only 1% of the light the host star sheds on it but NASA's Kepler Space Telescope managed to catch a glimpse of it in 2011. With mass a little over that of Jupiter, this 'hot Jupiter', being gravitationally tightly bound to the host, completes its orbit around the star in just about two and a half days! Its atmosphere boasts of light absorbing components like vapors of sodium and potassium, and possibly titanium oxide but none can account for its nearly absent reflectivity!

HD 106906 b, the Runaway Oddball

HD 106906 b, with a mass of eleven Jupiters, is an exoplanet at a distance of 336 light years from Earth in the constellation of Crux. It orbits around its host spectroscopic binary star HD 106906 at a whopping distance of 738 AU, i.e. 110 billion kilometers! It is the most distant exoplanet discovered so far. It is way beyond the parent star's debris disk. HD 106906 b was first sighted with the Magellan telescope in Chile. But it took the Hubble Space Telescope to actually trace its motion after tireless precise measurements over 14 years! Orbit of HD 106906 b is extremely elongated and misaligned. Discovery of this exoplanet with its bizarre orbit has kindled the enthusiasm of solar planetary teams since it is a proxy to the much hypothesized features of the distant kindred of solar planets known as "Planet Nine". Further detailed study of HD 106906 b hopefully will reveal how these humongously massive planets form, in which time frame of the host star's formative years and how they end up in the current configuration!

WD1586 b, Exoplanet Larger than its Host Star

WD1586 b, the first Jupiter-sized exoplanet orbiting a much smaller white dwarf WD1586 barely 80 light years away from the Earth in the constellation, Draco was discovered by NASA's Goddard Space Center in September, 2020. NASA used its TESS facility along with the now retired Spitzer Space Telescope to spy on this exoplanet literally hugging its host star 1/7th its size! Astronomers are yet to understand how this giant exoplanet got so close to the 10 billion years old parent white dwarf. Typically during its formation, WD1856 should have engulfed all nearby celestial objects and incinerated them, including WD1586 b in its current position. The only plausible explanation is that WD1586 b must have formed at least fifty times farther away and slowly migrated inwards.

PSR J1719-1438 b, Exoplanet Compressed Out of a Star!

Discovery of the exoplanet PSR J1719-1438 b was announced in August 2011. Undoubtedly, this is one of the most exotic cosmic objects. It orbits around its host PSR J1719-1438, which is a pulsar which completes 10,000 rotations in a minute, i.e. with a spin period of 5.4 millisecond! PSR J1719-1438 b is larger than its host star. Its close orbit around the parent star would fit within the diameter of our Sun! PSR J1719-1438 b is only 40% the size of Jupiter but of the same mass as Jupiter. It is compressed so severely that it is surmised it could be a crystalline diamond five times as large as Earth. PSR J1719-1438 b must be the remnant of a star whose outer layers were blown off by the host pulsar. This probably accounts for its exquisite characteristics.

Bibliography

Adler, D. (2020). The Great Filter: A possible solution to the Fermi Paradox. https://astronomy.com/news/2020/11/the-great-filter-a-possible-solution-to-the-fermi-paradox.

Adams, B., White, A., Lenton, T. M. (2004). An analysis of some diverse approaches to modelling terrestrial net primary productivity. *Ecological Modelling* 177(3), 353-391. https://doi.org/10.1016/j.ecolmodel.2004.03.014.

Anglada-Escudé, Guillem, Tuomi, Mikko, Gerlach, Enrico, Barnes, Rory, Heller, René, Jenkins, James S., et al. (2013). A dynamically-packed planetary system around GJ 667C with three super-Earths in its habitable zone. *Astronomy & Astrophysics* 556, 126.

Armitage, P. (2008). Planetary formation and migration. *Scholarpedia* 3, 4479. doi: 10.4249/scholarpedia.4479.

Armstrong, J. C., Barnes, R., Domagal-Goldman, S., Breiner, J., Quinn, T. R., Meadows, V. S. (2014). Effects of extreme obliquity variations on the habitability of exoplanets. *Astrobiology* 14(4), 277-291. https://doi.org/10.1089/ast.2013.1129.

Barlow, N. (2014). *Mars: An Introduction to its Interior, Surface and Atmosphere. Cambridge Planetary Science*, (No. 8). Cambridge: Cambridge University Press, doi: http://dx.doi.org/10.1017/CBO9780511536069.

Brogi, M., Snellen, I. ~A. ~G., de Kok, R. ~J., Albrecht, S., Birkby, J., de Mooij, E. ~J. ~W., et al. 2012. The signature of orbital motion from the dayside of the planet τ. *Nature* 486, 502. doi: 10.1038/nature11161.

Barnes, R., Meadows, V. S., Evans, N. (2015). Comparative habitability of transiting exoplanets. *The Astrophysical Journal* 814(2), 91. https://doi.org/10.1088/0004-637X/814/2/91.

Baumann, H. (1922). Die anabiose der tardigraden. *Zoologische Jahrbücher* 45, 501-556.

Becquerel, P. (1950). La suspension de la vie au dessus de 1/20K absolu par demagnetization adiabatique de l'alun de fer dans le vide les plus eléve. *Comptes rendus de l'Académie des Sci.* 231, 261-263.

Belilla, J., Moreira, D., Jardillier, L., Reboul, G., Benzerara, K., López-García, J. M., et al. (2019). Hyperdiverse archaea near life limits at the polyextreme geothermal Dallol area. *Nature Ecology & Evolution* 3(11), 1552-1561. https://doi.org/10.1038/s41559-019-1005-0.

Bessel, F. W. (1844). XI. Observations of altitude and azimuth of the great comet of 1843, made at St. Helena. *Monthly Notices of the Royal Astronomical Society* 6, 136. doi: 10.1093/mnras/6.11.136.

Bethe, H.A. (1938). Energy production in stars. *Physical Review* 55, 434.

Beust, H., Vidal-Madjar, A., Ferlet, R., Lagrange-Henri, A. -M. (1990). The inner part of the protoplanetary disk around β Pictoris. *ESA Special Publication* 315, 81.

BIOKon In Space (BIOKIS). *NASA*. 17 May 2011. Retrieved 2011-05-24.

Bloom, S. A. (1981). Similarity indices in community studies: potential pitfalls. *Marine Ecology Progress Series* 5, 125.

BOSS on EXPOSE-R2-Comparative Investigations on Biofilm and Planktonic cells of Deinococcus geothermalis as Mission Preparation Tests. EPSC Abstracts. Vol. 8, EPSC2013-930, 2013. European Planetary Science Congress 2013.

Boyle, T. P., Smillie, G. M., Anderson, J. C., Beeson, D. R. (1990). A sensitivity analysis of nine diversity and seven similarity indices. *Research Journal of the Water Pollution Control Federation* 62(6), 749-762. JSTOR.

Brandstätter, F. (2008). Mineralogical alteration of artificial meteorites during atmospheric entry. The STONE-5 experiment. *Planetary and Space Science* 56(7), 976-984. doi: 10.1016/j.pss.2007.12.014.

Brandt, Annette, De Vera, Jean-Pierre, Onofri, Silvano, Ott, Sieglinde (2014). Viability of the lichen Xanthoria elegans and its symbionts after 18 months of space exposure and simulated Mars conditions on the ISS. *International Journal of Astrobiology* 14(3), 411-425. Bibcode: 2015IJAsB..14..411B. doi: 10.1017/S1473550414000214.

Bray, J. R. and Curtis, J. T. (1957). An ordination of the upland forest communities of Southern Wisconsin. *Ecological Monographs* 27, 325.

Brennard, Emma. (2011). *Tardigrades: Water bears in space. BBC.* Retrieved 2011-05-24.

Britannica, The Editors of Encyclopaedia. "science". Encyclopedia Britannica, 31 Dec. 2020, https://www.britannica.com/science/science. Accessed 23 January 2022.

Budyko (Ed.). (1974). *Climate and Life* (English ed. edition). New York: Academic Press.

Burchell, M. (2001). Survivability of bacteria in hypervelocity impact. *Icarus* 154(2), 545-547. doi: 10.1006/icar.2001.6738.

Carone, L., Keppens, R., Decin, L. (2016). Connecting the dots III: Night side cooling and surface friction affect climates of tidally locked terrestrial planets. *Monthly Notices of the Royal Astronomical Society* 461, 385.

Catling, D. C., Krissansen-Totton, J., Kiang, N. Y., Crisp, D., Robinson, T. D., DasSarma, S. (2018). Exoplanet Biosignatures: A Framework for Their Assessment. *Astrobiology* 18(6), 709-738. https://doi.org/10.1089/ast.2017.1737.

Cha, S. H. (2007). Comprehensive survey on distance/similarity measures between probability density functions. *International Journal of Mathematical Models and Methods in Applied Sciences* 1, 300.

Chen, H., Wolf, E. T., Zhan, Z., Horton, D. E. (2019). Habitability and spectroscopic observability of warm M-dwarf exoplanets evaluated with a 3D chemistry-climate model. *The Astrophysical Journal* 886(1), 16. https://doi.org/10.3847/1538-4357/ab4f7e.

Christopher, C. C., Yung, Y. L., Liang, M. C., Parkinson, C. D., Liang, M. –C., Hartman, H., et al. (2006). *Enceladus: Cassini Observations and Implications for the Search for Life.* American Geophysical Union, Fall Meeting 2006, abstract id. P13B-0176.

Clarke, A. (2017). *The Metabolic Theory of Ecology.* Principles of Thermal Ecology: Temperature, Energy, and Life, Oxford University Press. https://www.oxfordscholarship.com/view/10.1093/oso/9780199551668.001.0001/oso-9780199551668-chapter-12.

Clarke, K. R. (1993). Non-parametric multivariate analyses of changes in community structure. *Australian Journal of Ecology* 18(1), 117-143. https://doi.org/10.1111/j.1442-9993.1993.tb00438.x.

Claudia Pacelli, Laura Selbmann, Laura Zucconi, Jean-Pierre De Vera, Elke Rabbow, Gerda Horneck, et al. (2017). BIOMEX Experiment: Ultrastructural alterations, molecular damage and survival of the fungus cryomyces antarcticus after the experiment verification tests. *Origins of Life and Evolution of Biospheres* 47(2): 187-202.

Clegg, J. S. (1973). Do dried cryptobiotes have a metabolism? pp. 141-147. *In*: Crowe, John H., Madin, K. A. [eds.]. *Anhydrobiosis*. Dowden, Hutchinson and Ross, Stroudsburg.

Cockell, C. ~S., Brandt, D., Hand, K., Lee, P. (2001). Microbial Mats of the Tswaing Impact Crater: Results of a South African Exobiology Expedition and Implications for the Search for Biological Molecules on Mars, Lunar and Planetary Science Conference.

Cockell, C. S., Rettberg, Petra, Rabbow, Elke; Olson-Francis, Karen (2011). Exposure of phototrophs to 548 days in low Earth orbit: microbial selection pressures in outer space and on early earth. *The ISME Journal* 5(10), 1671-1682. doi: 10.1038/ismej.2011.46. PMC 3176519. PMID 21593797.

Cockell, C. S., Bush, T., Bryce, C., Direito, S., Fox-Powell, M., Harrison, J. P., et al. (2016). Habitability: A review. *Astrobiology* 16(1), 89-117. https://doi.org/10.1089/ast.2015.1295.

Cockell, C. S., Stevens, A. H., Prescott, R. (2019). Habitability is a binary property. *Nature Astronomy* 3(11), 956-957. https://doi.org/10.1038/s41550-019-0916-7.

Cowan, N. B., Abbot, D. S. (2014). Water cycling between ocean and mantle: Super-earths need not be waterworlds. *The Astrophysical Journal* 781, 27. doi: 10.1088/0004-637X/781/1/27.

Cramer, W., Kicklighter, D. W., Bondeau, A., Iii, B. M., Churkina, G., Nemry, B., et al. (1999). Comparing global models of terrestrial net primary productivity (NPP): Overview and key results. *Global Change Biology* 5(S1), 1-15. https://doi.org/10.1046/j.1365-2486.1999.00009.x.

Carlson, B., Lacis, A., Colose, C., Marshak, A., Su, W., Lorentz, S. et al. (2019). Spectral signature of the biosphere: NISTAR finds it in our solar system from the lagrangian L-1 point. *Geophysical Research Letters* 46(10), 679. doi: 10.1029/2019GL083736.

Cuntz, M., Guinan, E. F. (2016). About exobiology: The case for dwarf K stars. *The Astrophysical Journal* 827, 79. doi: 10.3847/0004-637X/827/1/79.

Dai, X., Guerras, E. (2018). Probing planets in extragalactic galaxies using quasar microlensing. *The Astrophysical Journal Letters* 853, L27. doi: 10.3847/2041-8213/aaa5fb.

Dawson, R. I. (2018). haex.book. 114. doi: 10.1007/978-3-319-55333-714.

Degma, P., Guidetti, R. (2007). Notes to the current checklist of Tardigrada. *Zootaxa* 1579, 41-53.

Degma, P., Bertolani, R., Guidetti, R. (2009). Actual checklist of Tardigrada species. (Ver. 33: 15-10-2017). http://www.tardigrada.modena.unimo.it/miscellanea/Actual%20checklist%20of%20Tardigrada.pdf.

de La Torre Noetzel, R. (2007). BIOPAN experiment LICHENS on the Foton M2 mission: Pre-flight verification tests of the Rhizocarpon geographicum-granite ecosystem. *Advances in Space Research* 40(11), 1665-1671. doi: 10.1016/j.asr.2007.02.022.

De La Vega, U. P., Rettberg, P., Reitz, G. (2007). Simulation of the environmental climate conditions on martian surface and its effect on Deinococcus radiodurans. *Advances in Space Research* 40(11), 1672-1677. Bibcode: 2007AdSpR..40.1672D. doi: 10.1016/j.asr.2007.05.022.

Demory, Brice-Olivier, Gillon, Michael, Madhusudhan, Nikku, Queloz, Didier. (2015). Variability in the super-Earth 55 Cnc e. *Monthly Notices of the Royal Astronomical Society* 455(2), 2018-2027. arXiv: 1505.00269. Bibcode: 2016MNRAS.455.2018D. doi: 10.1093/mnras/stv2239. S2CID 53662519.

de Vera, J. P., Horneck, G., Rettberg, P., Ott, S. (2004). The potential of the lichen symbiosis to cope with the extreme conditions of outer space II: Germination capacity of lichen ascospores in response to simulated space conditions. *Advances in Space Research* 33(8), 1236-1243. Bibcode: 2004AdSpR..33.1236D. doi: 10.1016/j.asr.2003.10.035. PMID 15806704.

de Vera, J. P., Dulai, S., Kereszturi, A., Koncz, L., Pocs, T. (2013). Results on the survival of cryptobiotic cyanobacteria samples after exposure to Mars-like environmental conditions. *International Journal of Astrobiology* 13(1), 35-44. Bibcode: 2014IJAsB..13...35D. doi: 10.1017/S1473550413000323.

Deza, E. and Deza, M. (2006). Dictionary of distances. Elsevier. Reviewed in the Newsletter. *Journal of the European Mathematical Society* 64, 57.

Dobrovolskis, A. R. (2013). Insolation on exoplanets with eccentricity and obliquity. *Icarus*, 226, 760. doi: 10.1016/j.icarus.2013.06.026.

Dobrovolskis, A. R. (2015). Insolation patterns on eccentric exoplanets. *Icarus* 250, 395. doi: 10.1016/j.icarus.2014.12.017.

Dose, K. (1995). ERA-experiment "space biochemistry". *Advances in Space Research* 16(8), 119-129. doi: 10.1016/0273-1177(95)00280-R.

Drake, F., Sobel, D. (1992). *Is Anyone Out There? The Scientific Search for Extraterrestrial Intelligence.* New York: Delacorte Press. pp. 55-62. ISBN: 0-385-31122-2.

Dublin, M., Volz, P. ~A. (1973). Space-related research in mycology concurrent with the first decade of manned space exploration. *Space Life Sciences* 4, 223. doi: 10.1007/BF00924469.

Elkins-Tanton, Linda T., Seager, Sara (2008). *The Astrophysical Journal* 688, 628.

Evans, T. M., Pont, F. D. R., Sing, D. K., Aigrain, S., Barstow, J. K., Désert, J. M., et al. (2013). The deep blue color of HD189733b: Albedo measurements with hubble space telescope/space telescope imaging spectrograph at visible wavelengths. *The Astrophysical Journal* 772(2), L16. arXiv: 1307.3239. Bibcode: 2013ApJ...772L..16E. doi: 10.1088/2041-8205/772/2/L16. S2CID 38344760.

Expose-R: Exposure of Osmophilic Microbes to Space Environment. *NASA*. 26 April 2013. Retrieved 2013-08-07.

Fabrycky, D., Tremaine, S. (2007). *The Astrophysical Journal* 669, 1298. doi: 10.1086/521702.

Fajardo-Cavazos, P., Link, L., Melosh, H. J., Nicholson, W. L. (2005). Bacillus subtilis spores on artificial meteorites survive hypervelocity atmospheric entry: Implications for lithopanspermia. *Astrobiology* 5(6), 726-736. Bibcode: 2005AsBio...5..726F. doi: 10.1089/ast.2005.5.726. PMID 16379527.

Ferlet, R., Hobbs, L. M., Madjar, A. V. (1987). *AA*. 185, 267.

Freedman, R., Kaufmann, W. J. (2007). *Universe*. USA. ISBN-13: 9780716785842.

Förster, F., Liang, C., Shkumatov, A., Beisser, D., Engelmann, J. C., Schnölzer, M., et al. (2009). Tardigrade workbench: comparing stress related proteins, sequence-similar and functional protein clusters as well as RNA elements in tardigrades. *BMC Genomics* 10, 469. pmid: 19821996. doi: 10.1186/471-2164-10-469.

Gaidos, E., Deschenes, B., Dundon, L., Fagan, K., Menviel-Hessler, L., Moskovitz, N., et al. (2005). Beyond the principle of plentitude: A review of terrestrial planet habitability. *Astrobiology* 5, 100. doi: 10.1089/ast.2005.5.100.

Gastine, T., Wicht, J., Duarte, L. D. V., Heimpel, M., Becker, A. (2014). Explaining Jupiter's magnetic field and equatorial jet dynamics. *Geophysical Research Letters* 41, 5410. doi: 10.1002/2014GL060814.

Gorshkov, V. G., Makarieva, A. M., Gorshkov, V. V. (2000). *Biotic Regulation of the Environment: Key Issues of Global Change*. Springer Science & Business Media.

Gorshkov, V. G., Makarieva, A. M., Gorshkov, V. V. (2004). Revising the fundamentals of ecological knowledge: The biota-environment interaction. *Ecological Complexity* 1(1), 17-36. https://doi.org/10.1016/j.ecocom.2003.09.002.

Greaves, Jane S., Bains, William, Petkowski, Janusz J., Seager, Sara, Sousa-Silva, Clara, Ranjan, Sukrit, et al. (2020). arXiv: 2012.05844.

Greenacre, M. and Primicerio, R. (2013). *Multivariate Analysis of Ecological Data*. Published by: Fundaci{o}n BBVA.

Grigoryev, Y. G. (1972). Influence of Cosmos 368 space flight conditions on radiation effects in yeasts, hydrogen bacteria and seeds of lettuce and pea. *Life Sciences and Space Research* 10, 113-118. PMID 11898831.

Grohme, M. A., Mali, B., Wełnicz, W., Michel, S., Schill, R. O., Frohme, M. (2013). The aquaporin channel repertoire of the tardigrade Milnesium tardigradum. *Bioinformatics and Biology Insights* 7, 153-165. https://doi.org/10.4137/BBI.S11497.

Guidetti, R., Bertolani, R. (2005). Tardigrade taxonomy: an updated checklist of the taxa and a list of characters for their identification. *Zootaxa* 845, 1-46.

Guidetti, R., Altiero, T., Rebecchi, L. (2011). On dormancy strategies in tardigrades. *Journal of Insect Physiology* 57, 567-576.

Guidetti, R., Rizzo, A.M., Altiero, T., Rebecchi, L. (2012). What can we learn from the toughest animals of the Earth? Water bears (tardigrades) as multicellular model organisms in order to perform scientific preparations for lunar exploration. *Planetary and Space Science* 74, 97-102.

Häder, Donat-P., Richter, Peter R., Strauch, S. M., Schuster, M. (2006). Aquacells — Flagellates under long-term microgravity and potential usage for life support systems. *Microgravity Science and Technology* 18(210): 210-214. doi: 10.1007/BF02870411.

Hagen, C. A., Hawrylewicz, E. J., Ehrlich, R. (1967). Survival of microorganisms in a simulated martian environment: II. Moisture and oxygen requirements for germination of bacillus cereus and bacillus subtilis var. Niger spores. *Applied Microbiology* 15(2), 285-291. doi: 10.1128/AEM.15.2.285-291.1967. PMC 546892. PMID 4961769.

Hamblin, W. K., Christiansen, E. H. (2003). *Earth's Dynamic Systems*. Pearson, India, 10th Edition, ISBN-13: 978-0131420663.

Harrison, J. P., Gheeraert, N., Tsigelnitskiy, D., Cockell, C. S. (2013). The limits for life under multiple extremes. *Trends in Microbiology* 21(4), 204-212. https://doi.org/10.1016/j.tim.2013.01.006.

Hart, M. H. (1978). The evolution of the atmosphere of the earth. *Icarus* 33, 23-39. https://doi.org/10.1016/0019-1035(78)90021-0.

Hashimoto, T., Horikawa, D. D., Saito, Y., Kuwahara, H., Kozuka-Hata, H., Shin-I, T. et al. (2016). Extremotolerant tardigrade genome and improved radiotolerance of human cultured cells by tardigrade-unique protein. *Nature Communications* 7, 12808.

Hawrylewicz, E., Gowdy, B., Ehrlich, R. (1962). Micro-organisms under a simulated martian environment. *Nature* 193(4814): 497. doi: 10.1038/193497a0.

Hawrylewicz, E., Hagen, C. A., Tolkacz, V., Anderson, B. T., Ewing, M. (1968). Probability of growth pG of viable microorganisms in Martian environments. *Life Sciences and Space Research VI.* pp. 146-156.

Hegde, S., Kaltenegger, L. (2013). Colors of extreme exo-earth environments. *Astrobiology* 13, 47. doi: 10.1089/ast.2012.0849.

Helled R., Vazan A. (2013). The evolution and survival of gaseous protoplanets embedded in a disk. *AAS/Division for Planetary Sciences Meeting Abstracts* #45.

Heller, R., Armstrong, J. (2014). Superhabitable worlds. *Astrobiology* 14(1), 50-66. https://doi.org/10.1089/ast.2013.1088.

Helled, R., Bodenheimer, P., Podolak, M., Boley, A., Meru, F., Nayakshin, S., et al. (2014). Giant planet formation, evolution, and internal structure. *In*: Beuther, Henrik, Klessen, Ralf S., Dullemond, Cornelis P., Henning, Thomas. [eds.]. *Protostars and Planets* VI, 643. doi: 10.2458/azu_uapress_9780816531240-ch028.

Heller, R. (2020). Habitability is a continuous property of nature. *Nature Astronomy* 4(4), 294-295. https://doi.org/10.1038/s41550-020-1063-x.

Hengherr, S., Heyer, A. G., Köhler, H. R., Schill, R. O. (2008). Trehalose and anhydrobiosis in tardigrades-evidence for divergence in response to dehydration. *The FEBS Journal* 275, 281-288.

Hengherr, S., Worland, M.R., Reuner, A., Brummer, F., Schill, R.O. (2009). Freeze tolerance, supercooling points and ice formation: comparative studies on the subzero temperature survival of limno-terrestrial tardigrades. *The Journal of Experimental Biology* 212, 802-807.

Hickman, Leo. (2010). Stephen hawking takes a hard line on aliens. The Guardian. Retrieved 24 February 2012.

Higashibata, A. (2006). Decreased expression of myogenic transcription factors and myosin heavy chains in Caenorhabditis elegans muscles developed during spaceflight. *Journal of Experimental Biology* 209(16), 3209-3218. doi: 10.1242/jeb.02365. PMID 16888068.

Hoeijmakers, H. ~J., Ehrenreich, D., Kitzmann, D., Allart, R., Grimm, S. ~L., Seidel, J. ~V., et al. 2019. A spectral survey of an ultra-Hot Jupiter. Detection of metals in the transmission spectrum of KELT-9 b. *Astronomy and Astrophysics* 627, A165. doi: 10.1051/0004-6361/201935089.

Horikawa, D. D., Sakashita, T., Katagiri, C., Watanabe, M., Kikawada, T., Nakahara, Y. et al. (2006). Radiation tolerance in the tardigrade Milnesium tardigradum. *International Journal of Radiation Biology* 82, 843-848.

Horneck, G. (2008). Microbial rock inhabitants survive hypervelocity impacts on Mars-like host planets: First phase of lithopanspermia experimentally Tested. *Astrobiology* 8(1), 17-44. Bibcode: 2008AsBio...8...17H. doi: 10.1089/ast.2007.0134. PMID 18237257.

Horneck, G. (2012). Resistance of bacterial endospores to outer space for planetary protection purposes—experiment PROTECT of the EXPOSE-E Mission. *Astrobiology* 12(5): 445-456. doi: 10.1089/ast.2011.0737.

Hotchin, J., Lorenz, P., Hemenway, C. (1965). Survival of microorganisms in space. *Nature* 206(4983), 442-445. Bibcode: 1965Natur.206..442H. doi: 10.1038/206442a0. PMID 4284122. S2CID 4156325.

Hotchin, J. (1968). The microbiology of space. *Journal of the British Interplanetary Society* 21, 122. Bibcode: 1968JBIS...21..122H.

Howell, R. R., Rathbun, J. A., Spencer, J. R. (2018). *AAS/Division for Planetary Sciences Meeting Abstracts* #50.

Howell, S. ~M., Chou, L., Thompson, M., Bouchard, M. ~C., Cusson, S., Marcus, M. ~L., et al. 2018. Camilla: A centaur reconnaissance and impact mission concept. *Planetary Space Science Procspie Proceedings* 164, 184. doi: 10.1016/j.pss.2018.07.008.

Huang, S. -S. (1959). The problem of life in the universe and the mode of star formation. *Publications of the Astronomical Society of the Pacific* 71, 421. https://doi.org/10.1086/127417.

Häder, D. P., Richter, P., Strauch, S., Schuster, M. (2006). Aquacells — Flagellates under long-term microgravity and potential usage for life support systems. *Microgravity Science and Technology* 18(210), 210-214. Bibcode: 2006MicST..18..210H. doi: 10.1007/BF02870411. S2CID 121659796.

Imshenetskiĭ, A. A., Kuzyurina, L. A., Yakshina, V. M. (1979). Xerophytic microorganisms multiplying under conditions close to Martian ones. *Mikrobiologiia* 48(1), 76-79. PMID 106224.

Imshenetskiĭ, A. A., Murzakov, B. G., Evdokimova, M. D., Dorofeeva, I. K. (1984). Survival of bacteria in the artificial mars unit. *Mikrobiologiia* 53(5), 731-737. PMID 6439981.

International Caenorhabditis elegans Experiment First Flight-Genomics (ICE-First-Genomics). November 22, 2016.

Isaacman, R., Sagan, C. (1977). Computer simulations of planetary accretion dynamics: Sensitivity to initial conditions. *Icarus* 31, 510. doi: 10.1016/0019-1035(77)90153-1.

Ito, A., Adachi, M., Yamagata, Y., Suzuki, R., Saigusa, N., Sekine, H. 2011. Evaluation of ecosystem services for good balance between climate change prevention and biodiversity conservation, AGU Fall Meeting Abstracts.

Jasechko, S., Sharp, Z. D., Gibson, J. J., Birks, S. J., Yi, Y., Fawcett, P. J. (2013). Terrestrial water fluxes dominated by transpiration. *Nature* 496, 347-350. https://doi.org/10.1038/nature11983.

JET: Journey to Enceladus and Titan, NASA, Sotin, C., Altwegg, K., Brown, R. H., Hand, K., Lunine, J. I., Soderblom, J., Spencer, Tortora, P. and the JET Team, 42nd Lunar and Planetary Science Conference, 2005.

Jones, B. (2008). Exoplanets search methods, discoveries, and prospects for astrobiology. *International Journal of Astrobiology* 7, 279.

Jänchen, Jochen, Feyh, Nina, Szewzyk, Ulrich, de Vera, Jean-Pierre P. (2015). Provision of water by halite deliquescence for Nostoc commune biofilms under Mars relevant surface conditions. *International Journal of Astrobiology* 15(2), 107-118. Bibcode: 2016IJAsB..15..107J. doi: 10.1017/S147355041500018X.

Joshi, A. W., Rana, N. C. (2011). *Our Solar System*. New Age International (P) Ltd, India. ISBN-13, 9788122431469.

Jönsson, K. I., Guidetti, R. (2001). Effects of methyl-bromide fumigation on anhydrobiotic micrometazoans. *Ecotoxicology and Environmental Safety* 50, 72-75.

Jönsson, K. I. (2007). Tardigrades as a potential model organism in space research. *Astrobiology* 7, 757-766.

Jönsson, K. I., Rabbow, E., Schill, R. O., Harms-Ringdahl, M., Rettberg, P. (2008). Tardigrades survive exposure to space in low Earth orbit. *Current Biology* 18(17), R729-R731. doi: 10.1016/j.cub.2008.06.048. PMID 18786368. S2CID 8566993.

Jönsson, K. Ingemar, Wojcik, Andrzej (2017). Tolerance to X-rays and Heavy Ions (Fe, He) in the Tardigrade Richtersius coronifer and the Bdelloid Rotifer Mniobia russeola. *Astrobiology* 17(2), 163-167. Bibcode: 2017AsBio..17..163J. doi: 10.1089/ast.2015.1462. ISSN 1531-1074. PMID 28206820.

Kaltenegger, L., Sasselov, D. (2011). Exploring the habitable zone for kepler planetary candidates. *The Astrophysical Journal Letters* 736, L25. https://doi.org/10.1088/2041-8205/736/2/L25.

Kane, S. R. (2013). The habitable zone: Basic concepts. pp. 3-12. *In*: J.-P. de Vera, J. Seckbach (eds.). *Habitability of Other Planets and Satellites*. Springer Netherlands. https://doi.org/10.1007/978-94-007-6546-7_1.

Kardashev, N. ~S., (1964). Transmission of information by extraterrestrial civilizations. *Sovast* 8, 217.

Kardashev, Nikolai. (1985). On the inevitability and the possible structures of supercivilizations. pp. 497-504. *In*: Dordrecht, D. Reidel (ed.). *The Search for Extraterrestrial Life: Recent Developments*. Proceedings of the Symposium, Boston, MA, June 18-21, 1984 (A86-38126 17-88). Publishing Co.

Karttunen, H., Kröger, P., Oja, H., Poutanen, M., Donner, K. J. (2007). *Fundamental Astronomy*. 5th ed., Vol. XI, 510, Springer-Verlag Berlin Heidelberg. 10.1007/978-3-540-34144-4.

Kashyap J. M., Gudennavar, S. B., Doshi, U., Safonova, M. (2017). Indexing of exoplanets in search for potential habitability: Application to Mars-like worlds. *Astrophysics and Space Science* 362(8), 146. https://doi.org/10.1007/s10509-017-3131-y.

Kashyap, J. M., Roszkowska, M., Kaczmarek, Ł. (2018). Tardigrade indexing approach on exoplanets. *Life Sciences in Space Research* 19, 13-16. doi: 10.1016/j.lssr.2018.08.001.

Kashyap, Jagadeesh, M., Rao Valluri, S., Kari, V., Kubska, K., Kaczmarek, Ł. (2020). Indexing Exoplanets with Physical Conditions Potentially Suitable for Rock-Dependent Extremophiles. *Life* 10, 10. https://doi.org/10.3390/life10020010.

Kasting, J. F., Whitmire, D. P., Reynolds, R. T. (1993). Habitable zones around main sequence stars. *Icarus* 101, 108. doi: 10.1006/icar.1993.1010.

Kaufman, M. (2019). The interiors of exoplanets may well hold the key to their Habitability. https://manyworlds.space/.

Kawabe, R., Ishiguro, M., Omodaka, T., Kitamura, Y., Miyama, S. -M. (1993). Discovery of a rotating protoplanetary gas disk around the young star GG Tauri. *The Astrophysical Journal* 404, L63. doi: 10.1086/186744.

Kawaguchi, Yuko, Hashimoto, Hirofumi, Yokobori, Shin-ichi, Yamagishi, Akihiko, Shibuya, Mio, Kinoshita, Iori. (2018). Survival and DNA damage of cell-aggregate of Deinococcus spp. exposed to space for two-years in Tanpopo mission. 42nd COSPAR Scientific Assembly. Held 14-22 July 2018, in Pasadena, California, USA, Abstract id. F3.1-5-18. July.

Kindt, R. and Coe, R. (2005). Tree diversity analysis. A manual and software for common statistical methods for ecological and biodiversity studies ICRAF.

Kislyakova, Kristina, G. Holmströmhelmut, Mats, Lammerpetra, Odertand Maxim, Khodachenko, L. (2014). Magnetic moment and plasma environment of HD 209458b as determined from Lyα observations. *Science* 346, 981. doi: 10.1126/science.1257829.

Kleidon, A. (2012). How does the Earth system generate and maintain thermodynamic disequilibrium and what does it imply for the future of the planet? *Philosophical Transactions of the Royal Society A: Mathematical, Physical and Engineering Sciences* 370(1962), 1012-1040. https://doi.org/10.1098/rsta.2011.0316.

Klementiev, Konstantin E., Maksimov, Eugene G., Gvozdev, Danil A., Tsoraev, Georgy V., Protopopov, Fedor F., Elanskaya, Irina V., et al. (2019). Radioprotective role of cyanobacterial phycobilisomes. *Biochimica et Biophysica Acta (BBA) - Bioenergetics* 1860(2), 121-128. Bibcode: 2019BBAB.1860..121K. doi: 10.1016/j.bbabio.2018.11.018. PMID 30465750.

Koike, J., Oshima, T., Kobayashi, K., Kawasaki, Y. (1995). Studies in the search for life on Mars. *Advances in Space Research* 15(3), 211-214. Bibcode: 1995AdSpR..15..211K. doi: 10.1016/S0273-1177(99)80086-6. PMID 11539227.

Koike, J. (1996). Fundamental studies concerning planetary quarantine in space. *Advances in Space Research* 18(1-2), 339-344. Bibcode: 1996AdSpR..18a.339K. doi: 10.1016/0273-1177(95)00825-Y. PMID 11538982.

Komacek, T. D., Fauchez, T. J., Wolf, E. T., Abbot, D. S. (2020). Clouds will likely prevent the detection of water vapor in JWST transmission spectra of terrestrial exoplanets. *The Astrophysical Journal* 888(2), L20. https://doi.org/10.3847/2041-8213/ab6200.

Kopparapu, R. K., Ramirez, R., Kasting, J. F., Eymet, V., Robinson, T. D., Mahadevan, S., et al. (2013). Habitable zones around main-sequence stars: New estimates. *The Astrophysical Journal* 765, 131. https://doi.org/10.1088/0004-637X/765/2/131.

Kopparapu, R. K., Ramirez, R. M., SchottelKotte, J., Kasting, J. F., Domagal-Goldman, S. and Eymet, V. (2014). Habitable zones around main-sequence stars: dependence on planetary mass. *The Astrophysical Journal Letters* 787, L29. https://doi.org/10.1088/2041-8205/787/2/L29.

Krauthammer, Charles (2011). Are we alone in the universe? The Washington Post. Retrieved January 6, 2015.

Lemarchand, Guillermo A. (1994). *Detectability of Extraterrestrial Technological Activities*. Coseti. Republished in SETIQuest, Volume 1, Number 1, pp. 3-13.

Lenton, T. M. (1998). Gaia and natural selection. *Nature* 394(6692), 439-447. https://doi.org/10.1038/28792.

Looman, J. and Campbell, J. B. (1960). Adaptation of Sorensen's K (1948) for estimating unit affinities in prairie vegetation. *Ecology* 41, 409.

Lorenz, R. D. (2020). Maunder's work on planetary habitability in 1913: Early use of the term "Habitable Zone" and a "Drake Equation" calculation. *Research Notes of the AAS*, 4(6), 79. https://doi.org/10.3847/2515-5172/ab9831.

Lynch, M. D. J., Neufeld, J. D. (2015). Ecology and exploration of the rare biosphere. *Nature Reviews Microbiology* 13(4), 217-229. https://doi.org/10.1038/nrmicro3400.

Madhusudhan, N. (2019). Exoplanetary atmospheres: Key insights, challenges, and prospects. *Annual Review of Astronomy and Astrophysics* 57, 617. doi: 10.1146/*annurev-astro*-081817-051846.

Mamajek, E. E., Quillen, A. C., Pecaut, M. J., Moolekamp, F., Scott, E. L., Kenworthy M. A., et al. (2012). Planetary construction zones in occultation: Discovery of an extrasolar ring system transiting a young Sun-like star and future prospects for detecting eclipses by circumsecondary and circumplanetary Disks. *The Astronomical Journal* 143(3), 72. arXiv: 1108.4070. Bibcode: 2012AJ....143...72M. doi: 10.1088/0004-6256/143/3/72. S2CID 55818711.

Mancinelli, R. L., White, M. R., Rothschild, L. J. (1998). Biopan-survival I: Exposure of the osmophiles Synechococcus SP. (Nageli) and Haloarcula SP. To the space environment. *Advances in Space Research* 22(3), 327-334. Bibcode: 1998AdSpR..22..327M. doi: 10.1016/S0273-1177(98)00189-6.

Mancinelli, R. L. (2015). The affect [sic] of the space environment on the survival of Halorubrum chaoviator and Synechococcus (Nägeli): data from the Space Experiment OSMO on EXPOSE-R. *International Journal of Astrobiology* 14(Special Issue 1): 123-128. doi: 10.1017/S147355041400055X.

Mandushev, G., O'Donovan, F. T., Charbonneau, D., Torres, G., Latham, D. W., Bakos, G. A., et al. (2007). TrES-4: A transiting hot Jupiter of very low density. *The Astrophysical Journal* 667, L195. doi: 10.1086/522115.

Martiny, J. B. H., Bohannan, B. J. M., Brown, J. H., Colwell, R. K., Fuhrman, J. A., Green, J. L. (2006). Microbial biogeography: Putting microorganisms on the map. *Nature Reviews Microbiology* 4(2), 102-112. https://doi.org/10.1038/nrmicro1341.

Maunder, E. W. (1913). Are the planets inhabited? *London, New York, Harper & Brothers, 1913.* http://adsabs.harvard.edu/abs/1913api..book.....M.

Mayor, M., Queloz, D. (1995). A Jupiter-mass companion to a solar-type star. *Nature* 378, 355.

McInnes S. J., (1994). Zoogeographic distribution of terrestrial/freshwater tardigrades from current literature. *Journal of Natural History* 28, 257-352.

McKay, C. P. (2014). Requirements and limits for life in the context of exoplanets. *Proceedings of the National Academy of Sciences* 111(35), 12628-12633. https://doi.org/10.1073/pnas.1304212111.

Meeßen, J., Wuthenow, P., Schille, P., Rabbow, E., de Vera, J.-P.P. (2015). Resistance of the lichen buellia frigida to simulated space conditions during the preflight tests for BIOMEX — Viability assay and morphological stability. *Astrobiology* 15(8), 601-615. doi: 10.1089/ast.2015.1281.

Molina, R. D., Salazar, J. F., Martínez, J. A., Villegas, J. C., Arias, P. A. (2019). Forest-Induced Exponential Growth of Precipitation Along Climatological Wind Streamlines Over the Amazon. *Journal of Geophysical Research: Atmospheres* 124(5), 2589-2599. https://doi.org/10.1029/2018JD029534.

Moll, D. M., Vestal, J. R. (1992). Survival of microorganisms in smectite clays: Implications for martian exobiology. *Icarus* 98, 233. doi: 10.1016/0019-1035(92)90092-L.

Morozova, D., Möhlmann, D., Wagner, D. (2006). Survival of methanogenic archaea from Siberian permafrost under simulated martian thermal conditions (PDF). *Origins of Life and Evolution of Biospheres* 37(2), 189-200. Bibcode: 2007OLEB...37..189M. doi: 10.1007/s11084-006-9024-7. PMID 17160628. S2CID 15620946.

Močnik, T., Anderson, D. R., Brown, D. J. A., Collier Cameron, A., Delrez, L., Gillon, M., et al. (2016). *Publications of the Astronomical Society of the Pacific* 128, 124403. doi: 10.1088/1538-3873/128/970/124403.

Murray, C. D., Correia A. C. M. (2010). *Exoplanets* 15. arXiv: 1009.1738.

Meeßen, J., Wuthenow, P., Schille, P., Rabbow, E., de Vera, J. -P. P. (2015). Resistance of the lichen buellia frigida to simulated space conditions during the preflight tests for BIOMEX—viability assay and morphological stability. *Astrobiology* 15(8): 601-615. doi: 10.1089/ast.2015.1281.

Méndez, Abel, Rivera-Valentín, E. G. (2017). The Equilibrium Temperature of Planets in Elliptical Orbits. *The Astrophysical Journal* 837(1), L1. https://doi.org/10.3847/2041-8213/aa5f13.

Méndez, Abel, Rivera-Valentín, Edgard G., Schulze-Makuch, Dirk, Filiberto, Justin, Ramírez, Ramses M., Wood, Tana E., et al. (2021). Habitability models for astrobiology. *Astrobiology* 21(8), 1017-1027. doi: 10.1089/ast.2020.2342.

Nasir, A., Strauch, S.M., Becker, I., Sperling, A., Schuster, M., Richter P.R., et al. (2014). The influence of microgravity on Euglena gracilis as studied on Shenzhou 8. *Journal of Plant Biology* 16, 113-119. doi: 10.1111/plb.12067. PMID 23926886.

Nelson, D. R., Guidetti, R., Rebecchi, L. (2010). Chapter 14: *Tardigrada. Ecology and Classification of North American Freshwater Invertebrates*, third ed. Academic Press, San Diego, pp. 455-484.

Nelson D. R., Guidetti R., Rebecchi, L. (2015). Chapter 17: *Phylum Tardigrada. Ecology and General Biology: vol. 1: Thorp and Covich's Freshwater Invertebrates*. Elsevier. pp. 347-380.

Neuberger, Katja, Lux-Endrich, Astrid, Panitz, Corinna, Horneck, Gerda (2015). Survival of spores of Trichoderma longibrachiatum in space: data from the space experiment SPORES on EXPOSE-R. *International Journal of Astrobiology* 14 (Special Issue 1), 129-135. doi: 10.1017/S1473550414000408.

Nicholson, Wayne L., Krivushin, Kirill, Gilichinsky, U., Schuerger, Andrew C. (2012). Growth of Carnobacterium spp. from permafrost under low pressure, temperature, and anoxic atmosphere has implications for Earth microbes on Mars. *Proceedings of the National Academy of Sciences* 110(2), 666-671. Bibcode: 2013PNAS..110..666N. doi: 10.1073/pnas.1209793110. PMC 3545801. PMID 23267097.

Nordheim, T., Paranicas, C., Hand, K. ~P. (2017). Europa's surface radiation environment and considerations for *in-situ* sampling and biosignature detection, AGU Fall Meeting Abstracts, P52.

Novikova, N., Deshevaya, E., Levinskikh, M., Polikarpov, N., Poddubko, S. (2015). Study of the effects of the outer space environment on dormant forms of microorganisms, fungi and plants in the 'Expose-R' experiment. *International Journal of Astrobiology* 14(1), 137-142. doi: 10.1017/S1473550414000731.

Nowajewski, P., Rojas, M., Rojo, P., Kimeswenger, S. (2018). Atmospheric dynamics and habitability range in Earth-like aquaplanets obliquity simulations. *Icarus*. 305, 84-90. https://doi.org/10.1016/j.icarus.2018.01.002.

O'Callaghan, J. (2020). Life on Venus? Scientists hunt for the truth. *Nature* 586, 182-183. doi: https://doi.org/10.1038/d41586-020-02785-5.

Olsson-Francis, K., de la Torre, R., Towner, M. C., Cockell, C. S. (2009). Survival of akinetes (Resting-State Cells of Cyanobacteria) in low Earth orbit and simulated extraterrestrial conditions. *Origins of Life and Evolution of Biospheres* 39(6), 565-579. Bibcode: 2009OLEB...39..565O. doi: 10.1007/s11084-009-9167-4. PMID 19387863. S2CID 7228756.

Ono, F., Saigusa, M., Uozumi, T., Matsushima, Y., Ikeda, H., Saini, N. L. et al. (2008). Effect of high hydrostatic pressure on to life of the tiny animal tardigrade. *Journal of Physics and Chemistry of Solids* 69, 2297-2300.

Pan, Y., Cheong, C. M., Blevis, E. (2010). The climate change habitability index. *Interactions* 17(6), 29-33. https://doi.org/10.1145/1865245.1865253.

Pasini, D. L. S. et. al. (2013). EPSC2013, 396.

Pasini, D. L. S. et al. (2013). LPSC44, 1497.

Pasini D. L. S. et. al. (2014). EPSC2014, 67.

Pasini D. L. S. et al. (2014). LPSC45, 1789.

Pasini, J. L. S., Price, M. C. (2015). Panspermia survival scenarios for organisms that survive typical hypervelocity solar system impact events (PDF). *46th Lunar and Planetary Science Conference.*

Peacock, John, A. (1998), *Cosmological Physics.* Cambridge University Press. p. 66. ISBN 0-521-42270-1.

Pickering, E. C. (1890). Aid to astronomical research. *Nature* 43, 105. doi: 10.1038/043105b0.

Pigoń, A., Węglarska, B. (1955). Rate of metabolism in tardigrades during active life and anabiosis. *Nature* 176, 121-122.

Pikuta, E. V., Hoover, R. B., Tang, J. (2007). Microbial extremophiles at the limits of life. *Critical Reviews in Microbiology* 33(3), 183-209. https://doi.org/10.1080/10408410701451948.

Pogson, N. (1856). Magnitudes of thirty-six of the minor planets for the first day of each month of the year 1857. *Monthly Notices of the Royal Astronomical Society* 17, 12. doi: 10.1093/mnras/17.1.12.

Pollack J. B., Hubickyj O., Bodenheimer P., Lissauer J. J., Podolak M., Greenzweig Y. (1996). Formation of the giant planets by concurrent accretion of solids and gas. *Icarus* 124, 62. doi: 10.1006/icar.1996.0190.

Powell, Brian P., Kostov, Veselin B., Rappaport, Saul A., Borkovits, Tamas, Zasche, Petr, Tokovinin, Andrei, et al. (2021). A sextuply-eclipsing sextuple star System. arXiv: 2101.03433.

Radeloff, V. C., Dubinin, M., Coops, N. C., Allen, A. M., Brooks, T. M., Clayton, M. K. (2019). The Dynamic Habitat Indices (DHIs) from MODIS and global biodiversity. *Remote Sensing of Environment* 222, 204-214. https://doi.org/10.1016/j.rse.2018.12.009.

Raggio, J. (2011). Whole lichen thalli survive exposure to space conditions: results of lithopanspermia experiment with aspicilia fruticulosa. *Astrobiology* 11(4), 281-292. Bibcode: 2011AsBio..11..281R. doi: 10.1089/ast.2010.0588. PMID 21545267.

Raktim, Roy, Phani, Shilpa P., Sangram, Bagh (2016). A systems biology analysis unfolds the molecular pathways and networks of two proteobacteria in spaceflight and simulated microgravity conditions. *Astrobiology* 16(9), 677-689. Bibcode: 2016AsBio..16..677R. doi: 10.1089/ast.2015.1420. PMID 27623197.

Ramazzotti, G., Maucci, W. (1983). *Il Phylum Tardigrada 41. Memorie dellí Istituto Italiano di Idrobiologia*, Pallanza, pp. 1-1012.

Ramirez, R. M., Kaltenegger, L. (2017). A volcanic hydrogen habitable zone. *The Astrophysical Journal Letters* 837, L4. https://doi.org/10.3847/2041-8213/aa60c8.

Ramirez, R. M., Kaltenegger, L. (2018). A methane extension to the classical habitable zone. *The Astrophysical Journal* 858, 72. https://doi.org/10.3847/1538-4357/aab8fa.

Ramløv, H., Westh, P. (1992). Survival of the cryptobiotic eutardigrade Adorybiotus coronifer during cooling to minus 196°C: effect of cooling rate, trehalose level, and short-term acclimation. *Cryobiology* 29, 125-130.

Ramløv, H., Westh, P. (2001). Cryptobiosis in the eutardigrade Adorybiotus (Richtersius) coronifer: tolerance to alcohols, temperature and de novo protein synthesis. *Zoologischer Anzeiger* 240, 517-523.

Rana, N.C., Wilkinson, D.A. (1989). An empirical law of star formation in spiral galaxies. *Monthly Notices of the Royal Astronomical Society*, 231, 509-513.

Redd, S. C. (2016). *Masters Thesis*, 43.

Rizzo, A. M., Negroni, M., Altiero, T., Montorfano, G., Corsetto, P., Berselli, P., et al. (2010). Antioxidant defences in hydrated and desiccated states of the tardigrade Paramacrobiotus richtersi. *Comparative Biochemistry and Physiology B* 156, 115-121.

Roberts, T. L., Wynne, E. S. (1962). Studies with a simulated Martian environment. *Journal of the Astronautical Sciences* 10, 65-74.

Rodríguez-Mozos, J. ~M., Moya, A. (2017). *Monthly Notices of the Royal Astronomical Society* 471, 4628. doi: 10.1093/mnras/stx1910.

Rodríguez-López, L., Cardenas, R., Parra, O., González-Rodríguez, L., Martin, O., Urrutia, R. (2019). On the quantification of habitability: Merging the astrobiological and ecological schools. *International Journal of Astrobiology* 18(5), 412-415. https://doi.org/10.1017/S1473550418000344.

Rogers, L. A., Seager, S. (2010). Three possible origins for the gas layer on GJ 1214b. *The Astrophysical Journal* 716, 1208. doi: 10.1088/0004-637X/716/2/1208.

Ronan, C. A. (1964). *oach.book.*

Rosa, Zélia Miller Ana, Cubero Beatriz, Martín-Cerezo M. Luisa, Raguse Marina, Meeßen Joachim (2017). The effect of high-dose ionizing radiation on the astrobiological model lichen Circinaria gyrosa. *Astrobiology* 17(2), 145-153. Bibcode: 2017AsBio..17..145D. doi: 10.1089/ast.2015.1454. PMID 28206822.

Roten, C. A., Gallusser, A., Borruat, G. D., Udry, S. D., Karamata, D. (1998). Impact resistance of bacteria entrapped in small meteorites. *Bulletin de la Société Vaudoise des Sciences Naturelles* 86(1), 1-17.

Salazar, J. F., Poveda, G. (2009). Role of a simplified hydrological cycle and clouds in regulating the climate-biota system of Daisyworld. *Tellus B: Chemical and Physical Meteorology* 61(2), 483-497. https://doi.org/10.1111/j.1600-0889.2009.00411.x.

Sancho, L. G. (2007). Lichens survive in space: Results from the 2005 LICHENS experiment. *Astrobiology* 7(3), 443-454. Bibcode: 2007AsBio...7..443S. doi: 10.1089/ast.2006.0046. PMID 17630840.

Sarantopoulou, E., Gomoiu, I., Kollia, Z., Cefalas, A. C. (2011). Interplanetary survival probability of Aspergillus terreus spores under simulated solar vacuum ultraviolet irradiation. *Planetary and Space Science* 59(1), 63-78.

Sarantopoulou, E., Stefi, A., Kollia, Z., Palles, D., Petrou, P.S., Bourkoula, A., et al. (2014). Viability of Cladosporium herbarum spores under 157 nm laser and vacuum ultraviolet irradiation, low temperature (10 K) and vacuum. *Journal of Applied Physics* 116(10), 104701. Bibcode: 2014JAP...116j4701S. doi: 10.1063/1.4894621.

Schlichting, Hilke E., Chang, Philip. (2011). Warm saturns: On the nature of rings around extrasolar planets that reside inside the ice line. *The Astrophysical Journal* 734(2), 117. arXiv: 1104.3863. Bibcode: 2011ApJ...734..117S. doi: 10.1088/0004-637X/734/2/117. S2CID 42698264.

Schneider, J. 2011. Defining and cataloging exoplanets: The exoplanet.eu database, EPSC-DPS Joint Meeting 2011, 3.

Schramski, J. R., Dell, A. I., Grady, J. M., Sibly, R. M., Brown, J. H. (2015). Metabolic theory predicts whole-ecosystem properties. *Proceedings of the National Academy of Sciences of the United States of America* 112(8), 2617-2622. https://doi.org/10.1073/pnas.1423502112.

Schulz, J. (2007). *Bray-Curtis dissimilarity. Algorithms - Similarity.* Alfred-Wegener-Institute for Polar and Marine Research, Bremerhaven. Germany. http://www.code10.info/, Retrieved 01/06/2016.

Schulze-Makuch, D., Méndez, A., Fairén, A. G., von Paris, P., Turse, C., Boyer, G. et al. (2011). A two-tiered approach to assessing the habitability of exoplanets. *Astrobiology* 11(10), 1041-1052. https://doi.org/10.1089/ast.2010.0592.

Seager, S., Kuchner, M., Hier-Majumder, C. ~A., Militzer, B. (2007). Mass-radius relationships for solid exoplanets. *The Astrophysical Journal* 669(2), 1279-1297. doi: 10.1086/521346.

Seager, S., Deming, D. (2010). Exoplanets atmospheres. *Annu. Rev. Astron. Astrophys* 48, 631.

Seales, J., Lenardic, A. (2020). Uncertainty quantification in planetary thermal history models: Implications for hypotheses discrimination and habitability modeling. *The Astrophysical Journal* 893(2), 114. https://doi.org/10.3847/1538-4357/ab822b.

Seki, K., Toyoshima, M. (1998). Preserving tardigrades under pressure. *Nature* 395, 853-858.

Selsis, F., Kasting, J. F., Levrard, B., Paillet, J., Ribas, I., Delfosse, X. (2007). Habitable planets around the star Gliese 581? *Astronomy and Astrophysics* 476, 1373-1387. https://doi.org/10.1051/0004-6361:20078091.

Sengupta, S. (2016). The search for another earth. *Resonance* 21, 641.

Sholes, S. F., Krissansen-Totton, J., Catling, D.C. (2019). A maximum subsurface biomass on mars from untapped free energy: CO and H_2 as potential antibiosignatures. *Astrobiology* 19, 655-668. https://doi.org/10.1089/ast.2018.1835.

Showman, Adam P., Fortney, Jonathan J., Lewis, Nikole, K., Shabram, Megan (2013). Doppler signatures of the atmospheric circulation on Hot Jupiters. *The Astrophysical Journal* 762(1), 24. doi: 10.1088/0004-637X/762/1/24.

Silva, L., Vladilo, G., Schulte, P. M., Murante, G., Provenzale, A. (2017). From climate models to planetary habitability: Temperature constraints for complex life. *International Journal of Astrobiology* 16(3), 244-265. https://doi.org/10.1017/S1473550416000215.

Sotin, C., Jackson, J. -M., Seager, S. (2010). Terrestrial planet interiors. pp. 375-395. *In*: Seager, S. [ed.]. *Exoplanets*. doi: https://ui.adsabs.harvard.edu/abs/2010exop.book..375S}.

Spitoni, E., Gioannini, L., Matteucci, F. (2017). Galactic habitable zone around M and FGK stars with chemical evolution models that include dust. *Astronomy and Astrophysics* 605, A38. https://doi.org/10.1051/0004-6361/201730545.

Stan-Lotter, H. (2002). Astrobiology with haloarchaea from Permo-Triassic rock salt. *International Journal of Astrobiology* 1(4), 271-284. Bibcode: 2002IJAsB...1..271S. doi: 10.1017/S1473550403001307.

Stoker, C. R., Zent, A., Catling, D. C., Douglas, S., Marshall, J. R., Archer, D., et al. (2010). Habitability of the Phoenix landing site. *Journal of Geophysical Research: Planets* 115(E6). https://doi.org/10.1029/2009JE003421.

Strauch, S.M., Richter, P., Schuster, M., Häder, D.-P. (2010). The beating pattern of the flagellum of Euglena gracilis under altered gravity during parabolic flights. *Journal of Plant Physiology* 167(1), 41-46. doi: 10.1016/j.jplph.2009.07.009. PMID 19679374.

Strauch, Sebastian M., Becker, Ina, Pölloth, Laura, Richter, Peter, R., Haag, Ferdinand, W. M., Hauslage, Jens, et al. (2018). Restart capability of resting-states of Euglena gracilis after 9 months of dormancy: preparation for autonomous space flight experiments. *International Journal of Astrobiology* 17(2), 101-111. doi: 10.1017/S1473550417000131.

Strom, Paul A., Dennis Bodewits, Matthew M. Knight, Flavien Kiefer, Geraint H. Jones, Quentin Kral, et al. (2020). *Exocomets from a Solar System Perspective*. Publications of the Astronomical Society of the Pacific 132 101001.

Sánchez, Francisco Javier, Meeßen, Joachim, Ruiza, M. del Carmen, Sancho, Leopoldo G., de la Torre, Rosa (2013). UV-C tolerance of symbiotic Trebouxia sp. in the space-tested lichen species Rhizocarpon geographicum and Circinaria gyrosa: role of the hydration state and cortex/screening substances. *International Journal of Astrobiology* 13(1), 1-18. Bibcode: 2014IJAsB..13....1S. doi: 10.1017/S147355041300027X.

Thompson, J. R. (2018). *Europa' s Ocean Ascending*. https://europa.nasa.gov>news. 1977.

Taylor, G. R., Bailey, J. V., Benton, E. V. (1975). Physical dosimetric evaluations in the Apollo 16 microbial response experiment. *Life Sciences and Space Research* 13, 135-141. PMID 11913418.

Tecon, R., Or, D. (2017). Biophysical processes supporting the diversity of microbial life in soil. *FEMS Microbiology Reviews* 41(5), 599-623. https://doi.org/10.1093/femsre/fux039.

Tobie, Gabriel. (2015). Planetary science: Enceladus' hot springs. *Nature* 519(7542) 162-163.

Tung, H. C., Bramall, N. E. and Price, P. B. (2005). Microbial origin of excess methane in glacial ice and implications for life on Mars. *Proceedings of the National Academy of Sciences of the United States of America* 102, 18292.

Underwood, D. R., Jones, B. W., Sleep, P. N. (2003). The evolution of habitable zones during stellar lifetimes and its implications on the search for extraterrestrial life. *International Journal of Astrobiology* 2, 289-299. https://doi.org/10.1017/S1473550404001715.

Unsöld, A., Baschek, B. (2002). *The New Cosmos.* (5th ed.). Springer-Verlag Berlin Heidelberg. 10.1007/978-3-662-04356-1.

Van der Meer, J. (2006). Metabolic theories in ecology. *Trends in Ecology and Evolution* 21(3), 136-140. https://doi.org/10.1016/j.tree.2005.11.004.

Van Heck, H. J., Tackley, P. J. (2011). Coulomb stability of the 4π periodic Josephson effect of Majorana fermions. *Earth and Planetary Science Letters* 310, 252. doi: 10.1016/j.epsl.2011.07.029.

Ventrudo, B. (2016). The Armchair Astronomer, Nebulae, Mintaka Publishing, India, Vol. 1.

Walker, J. C. G., Hays, P. B., Kasting, J. F. (1981). A negative feedback mechanism for the long-term stabilization of Earth's surface temperature. *Journal of Geophysical Research: Oceans* 86(C10), 9776-9782. https://doi.org/10.1029/JC086iC10p09776.

Wall, Mike (2016). Fungi survive mars-like conditions on space station. *Space.com.* Retrieved 2016-01-29.

Wall, W. (2018). *A History of Optical Telescopes in Astronomy.* SpringerLink. ISBN: 978-3-319-99087-3.

Wang, Fischer 2015. Youth plus experience: the discovery of 51 Pegasi *b. European Physical Journal H,* doi: 10.1140/epjh/e2015-60041-5.

Wassmann, M. (2012). Survival of spores of the UV-resistant bacillus subtilis strain MW01 after exposure to low-earth orbit and simulated martian conditions: Data from the space experiment ADAPT on EXPOSE-E. *Astrobiology* 12(5), 498-507. doi: 10.1089/ast.2011.0772.

Watson, C. ~A., de Mooij, E. ~J. ~W., Steeghs, D., Marsh, T. ~R., Brogi, M., Gibson, N.~P., et al. (2019). Doppler tomography as a tool for detecting exoplanet atmospheres. *Monthly Notices of the Royal Astronomical Society* 490, 1991. doi: 10.1093/mnras/stz2679.

Weinberger, A. J., Becklin, E. E., Schneider, G., Smith, B. A., Lowrance, P. J., Silverstone, M. D., et al. (1999). The circumstellar disk of HD 141569 imaged with NICMOS. *The Astrophysical Journal* 525, L53. doi: 10.1086/312334.

Weizsacker, C. F. V. (1938). Ein Leben zwischen Physik und Philosophie Ino Weber *Physikalische Zeitschrift* 39, 633-645.

Wełnicz, W., Grohme, M. A., Kaczmarek, Ł., Schill, R. O., Frohme, M. (2011). Anhydrobiosis in tardigrades - the last decade. *Journal of Insect Physiology* 57, 577-583.

Williams, J. P., Hallsworth, J. E. (2009). Limits of life in hostile environments: No barriers to biosphere function? *Environmental Microbiology* 11(12), 3292-3308. https://doi.org/10.1111/j.1462-2920.2009.02079.x.

Willis, M., Ahrens, T., Bertani, L., Nash, C. (2006). Bugbuster—survivability of living bacteria upon shock compression. *Earth and Planetary Science Letters* 247(3-4), 185-196. Bibcode: 2006E&PSL.247..185W. doi: 10.1016/j.epsl.2006.03.054.

Winn, J. N., Fabricky, D., Albrecht, S., Johnson, J. A. (2010). Hot stars with Hot Jupiters have high obliquities. *The Astrophysical Journal Letters*. doi: 10.1088/2041-8205/718/2/L145.

Woolfson, M. M. (1978). A new tidal theory for the origin of the solar system. *Royal Astronomical Society, Quarterly Journal,* vol. 19. https://ui.adsabs.harvard.edu/abs/1978QJRAS..19..167W/abstract.

Wright, J. C. (2001). Cryptobiosis 300 years on from van Leeuwenhoek: what have we learned about tardigrades? *Zoologischer Anzeiger* 240, 563-582.

Wuchterl, G., Guillot, T., Lissauer, J. J. (2000). *Protostars and Planets* IV, 1081.

Wyatt, M. ~C. (2008). Evolution of debris disks. *Annual Review of Astronomy and Astrophysics* 46, 339-383.

Yamagishi, Akihiko, Kawaguchi, Yuko, Hashimoto, Hirofumi, Yano, Hajime, Imai, Eiichi, Kodaira, Satoshi, et al. (2018). Environmental data and survival data of deinococcus aetherius from the exposure facility of the Japan experimental module of the international space station obtained by the Tanpopo Mission. *Astrobiology* 18(11), 1369-1374. doi: 10.1089/ast.2017.1751.

Yamaguchi, A., Tanaka, S., Yamaguchi, S., Kuwahara, H., Takamura, C., Imajoh-Ohmi, S., et al. (2012). Two novel heat-soluble protein families abundantly expressed in an anhydrobiotic tardigrade. *PLoS One* 7, e44209. https://doi.org/10.1371/journal.pone.0044209.

Young, R. S., Deal, P. H., Bell, J., Allen, J. L. (1964). Bacteria under simulated Martian conditions. *Life Sciences and Space Research* 2, 105-111. PMID 11881642.

Zaks, D. P. M., Ramankutty, N., Barford, C. C., Foley, J. A. (2007). From Miami to Madison: Investigating the relationship between climate and terrestrial net primary production. *Global Biogeochemical Cycles* 21(3), GB3004. https://doi.org/10.1029/2006GB002705.

Zarka, P. 2010. Radioastronomy and the study of exoplanets. *Astronomical Society of the Pacific Conference Series* 430, 175.

Zhukova, A. I., Kondratyev, I. I. (1965). On artificial Martian conditions reproduced for microbiological research. *Life Sciences and Space Research* 3, 120-126.

Zimmermann, M. W., Gartenbach, K. E., Kranz, A. R. (1994). First radiobiological results of LDEF-1 experiment A0015 with Arabidopsis seed embryos and Sordaria fungus spores. *Advances in Space Research* 14(10), 47-51. Bibcode: 1994AdSpR..14...47Z. doi: 10.1016/0273-1177(94)90449-9. PMID 11539984.

Zsom, A. (2015). A population-based habitable zone perspective. *The Astrophysical Journal* 813(1), 9. https://doi.org/10.1088/0004-637X/813/1/9.

Zuluaga, J. I., Salazar, J. F., Cuartas-Restrepo, P., Poveda, G. (2014). The habitable zone of inhabited planets. *Biogeosciences Discussions* 11(6), 8443-8483. https://doi.org/10.5194/bgd-11-8443-2014.

Index

About the Authors

Madhu Kashyap Jagadeesh is an astrophysicist from Bengaluru, India. His passion towards astrophysics started during a Research Education Advancement Program (REAP), at Planetarium Bangalore. Later the passion towards exoplanets was developed at International Astronomical Youth Camp (IAYC), Germany. He has received the young scientist award from Indian Science Congress and a young researcher award from 51st ESLAB, ESTEC/European Space Agency, Netherlands. He has a teaching experience of more than 7 years as assistant professor at undergraduate Physics level. He has published many research articles in many reputed international journals on exoplanets and astrobiology. He is an expert panel member for many news channels in India.

Usha Shekhar is a retired Associate Professor of Physics from a reputed women's college in the state of Karnataka, India. She has to her credit an illustrious academic record. She obtained the Masters Degree in Physics topping the merit list as also scoring the highest percentage of marks among the students of the seven Science Departments of Bangalore University. She has a cache of gold and silver medals in recognition of her merit. In a teaching career spanning over 40 years, Usha Shekhar, with her vast stockpile of knowledge has propagated her passion for Physics amongst her students. Usha Shekhar is innovative in driving home seemingly difficult concepts in Physics by drawing examples from day to day life. She is an avid learner and keeps abreast of advancements in Physics and allied subjects. She has authored text books in Physics for college students and study material for aspirants of technical courses. She has been a member of several academic bodies. Since her retirement Usha Shekhar has been actively involved in writing content in Physics and topics of general scientific interest for reputed publishers.

www.ingramcontent.com/pod-product-compliance
Lightning Source LLC
Chambersburg PA
CBHW070730220326
41598CB00024BA/3374

* 9 7 8 1 0 3 2 2 5 1 3 5 6 *